T0092649

Studies in Big Data

Volume 32

Series editor

Janusz Kacprzyk, Polish Academy of Sciences, Warsaw, Poland
e-mail: kacprzyk@ibspan.waw.pl

About this Series

The series "Studies in Big Data" (SBD) publishes new developments and advances in the various areas of Big Data-quickly and with a high quality. The intent is to cover the theory, research, development, and applications of Big Data, as embedded in the fields of engineering, computer science, physics, economics and life sciences. The books of the series refer to the analysis and understanding of large, complex, and/or distributed data sets generated from recent digital sources coming from sensors or other physical instruments as well as simulations, crowd sourcing, social networks or other internet transactions, such as emails or video click streams and other. The series contains monographs, lecture notes and edited volumes in Big Data spanning the areas of computational intelligence incl. neural networks, evolutionary computation, soft computing, fuzzy systems, as well as artificial intelligence, data mining, modern statistics and Operations research, as well as self-organizing systems. Of particular value to both the contributors and the readership are the short publication timeframe and the world-wide distribution, which enable both wide and rapid dissemination of research output.

More information about this series at http://www.springer.com/series/11970

Tania Cerquitelli • Daniele Quercia • Frank Pasquale
Editors

Transparent Data Mining for Big and Small Data

Editors
Tania Cerquitelli
Department of Control
and Computer Engineering
Politecnico di Torino
Torino, Italy

Daniele Quercia
Bell Laboratories
Cambridge, UK

Frank Pasquale
Carey School of Law
University of Maryland
Baltimore, MD, USA

ISSN 2197-6503 ISSN 2197-6511 (electronic)
Studies in Big Data
ISBN 978-3-319-54023-8 ISBN 978-3-319-54024-5 (eBook)
DOI 10.1007/978-3-319-54024-5

Library of Congress Control Number: 2017936756

Printed on acid-free paper

This Springer imprint is published by Springer Nature
The registered company is Springer International Publishing AG
The registered company address is: Gewerbestrasse 11, 6330 Cham, Switzerland

Preface

Algorithms are increasingly impacting our lives. They promote healthy habits by recommending activities that minimize risks, facilitate financial transactions by estimating credit scores from multiple sources, and recommend what to buy by profiling purchasing patterns. They do all that based on data that is not only directly disclosed by people but also inferred from patterns of behavior and social networks.

Algorithms affect us, yet the processes behind them are hidden. They often work as black boxes. With little transparency, wrongdoing is possible. Algorithms could recommend activities that minimize health risks only for a subset of the population because of biased training data. They could perpetuate racial discrimination by refusing mortgages based on factors imperfectly tied to race. They could promote unfair price discrimination by offering higher online shopping prices to those who are able to pay them. Shrouded in secrecy and complexity, algorithmic decisions might well perpetuate bias and prejudice.

This book offers design principles for better algorithms. To ease readability, the book is divided into three parts, which are tailored to readers of different backgrounds. To ensure transparent mining, solutions should first and foremost increase transparency (Part I), plus they should not only be algorithmic (Part II) but also regulatory (Part III).

To begin with Part I, algorithms are increasingly used to make better decisions about public goods (e.g., health, safety, finance, employment), and requirements such as transparency and accountability are badly needed. In Chapter "The Tyranny of Data? The Bright and Dark Sides of Data-Driven Decision-Making for Social Good", Lepri et al. present some key ideas on how algorithms could meet those requirements without compromising predictive power. In times of "post-truth" politics—the political use of assertions that "feel true" but have no factual basis—also news media might benefit from transparency. Nowadays, algorithms are used to produce, distribute, and filter news articles. In Chapter "Enabling Accountability of Algorithmic Media: Transparency as a Constructive and Critical Lens", Diakopoulos introduces a model that enumerates different types of information that might be disclosed about such algorithms. In so doing, the model enables transparency and media accountability. More generally, to support transparency on the entire Web, the Princeton Web Transparency and Accountability Project (Chapter "The Princeton Web Transparency and Accountability Project") has continuously monitored thousands of web sites to uncover how user data is collected and used, potentially reducing information asymmetry.

Design principles for better algorithms are also of algorithmic nature, and that is why Part II focuses on algorithmic solutions. Datta et al. introduce a family of measures that quantify the degree of influence exerted by different input data on the output (Chapter "Algorithmic Transparency via Quantitative Input Influence"). These measures are called quantitative input influence (QII) measures and help identify discrimination and biases built in a variety of algorithms, including black-boxes ones (only full control of the input and full observability of the output are needed). But not all algorithms are black boxes. Rule-based classifiers could be easily interpreted by humans, yet they have been proven to be less accurate than state-of-the art algorithms. That is also because of ineffective traditional training methods. To partly fix that, in Chapter "Learning Interpretable Classification Rules with Boolean Compressed Sensing", Malioutov et al. propose new approaches for training Boolean rule-based classifiers. These approaches not only are well-grounded in theory but also have been shown to be accurate in practice. Still, the accuracy achieved by deep neural networks has been so far unbeaten. Huge amounts of training data are fed into an input layer of neurons, information is processed into a few (middle) hidden layers, and results come out of an output layer. To shed light on those hidden layers, visualization approaches of the inner functioning of neural networks have been recently proposed. Seifert et al. provide a comprehensive overview of these approaches, and they do so in the context of computer vision (Chapter "Visualizations of Deep Neural Networks in Computer Vision: A Survey").

Finally, Part III dwells on regulatory solutions that concern data release and processing—upon private data, models are created, and those models, in turn, produce algorithmic decisions. Here there are three steps. The first concerns data release. Current privacy regulations (including the "end-user license agreement") do not provide sufficient protection to individuals. Hutton and Henderson introduce new approaches for obtaining sustained and meaningful consent (Chapter "Beyond the EULA: Improving Consent for Data Mining"). The second step concerns data models. Despite being generated from private data, algorithm-generated models are not personal data in the strict meaning of law. To extend privacy protections to those emerging models, Giovanni Comandè proposes a new regulatory approach (Chapter "Regulating Algorithms' Regulation? First Ethico-Legal Principles, Problems, and Opportunities of Algorithms"). Finally, the third step concerns algorithmic decisions. In Chapter "What Role Can a Watchdog Organization Play in Ensuring Algorithmic Accountability?", AlgorithmWatch is presented. This is a watchdog and advocacy initiative that analyzes the effects of algorithmic decisions on human behavior and makes them more transparent and understandable.

There is huge potential for data mining in our society, but more transparency and accountability are needed. This book has introduced only a few of the encouraging initiatives that are beginning to emerge.

Torino, Italy,
Cambridge, UK
Baltimore, MD, USA
January 2017

Tania Cerquitelli
Daniele Quercia
Frank Pasquale

Contents

List of Contributors

Aisha Aamir Technische Universität Dresden, Dresden, Germany

Aparna Balagopalan Technische Universität Dresden, Dresden, Germany

Giovanni Comandé Scuola Superiore Sant'Anna Pisa, Pisa, Italy

Sanjeeb Dash IBM T. J. Watson Research Center, Yorktown Heights, NY, USA

Anupam Datta Carnegie Mellon University, Pittsburgh, PA, USA

Nicholas Diakopoulos Philip Merrill College of Journalism, University of Maryland, College Park, MD, USA

Amin Emad Institute for Genomic Biology, University of Illinois, Urbana Champaign, Urbana, IL, USA

1218 Thomas M. Siebel Center for Computer Science, University of Illinois, Urbana, IL, USA

Sebastian Grottel Technische Universität Dresden, Dresden, Germany

Stefan Gumhold Technische Universität Dresden, Dresden, Germany

Tristan Henderson School of Computer Science, University of St Andrews, St Andrews, UK

Luke Hutton Centre for Research in Computing, The Open University, Milton Keynes, UK

Dhruv Jain Technische Universität Dresden, Dresden, Germany

Bruno Lepri Fondazione Bruno Kessler, Trento Italy

Emmanuel Letouzé Data-Pop Alliance, New York, NY, USA

MIT Media Lab, Cambridge, MA, USA

Dmitry M. Malioutov IBM T. J. Watson Research Center, Yorktown Heights, NY, USA

Arvind Narayanan Princeton University, Princeton, NJ, USA

Nuria Oliver Data-Pop Alliance, New York, NY, USA

Dillon Reisman Princeton University, Princeton, NJ, USA

David Sangokoya Data-Pop Alliance, New York, NY, USA

Christin Seifert Technische Universität Dresden, Dresden, Germany

Shayak Sen Carnegie Mellon University, Pittsburgh, PA, USA

Abhinav Sharma Technische Universität Dresden, Dresden, Germany

Matthias Spielkamp AlgorithmWatch, Berlin, Germany

Jacopo Staiano Fortia Financial Solutions, Paris, France

Kush R. Varshney IBM T. J. Watson Research Center, Yorktown Heights, NY, USA

Yair Zick School of Computing, National University of Singapore, Singapore, Singapore

Acronyms

$\mathscr{A}$	Algorithm
$\mathscr{Q}_{\mathscr{A}}$	Quantity of interest
ι	Influence
1Rule	Boolean compressed sensing-based single rule learner
ADM	Automated decision making
AKI	Acute kidney injury
AI	Artificial intelligence
API	Application Program Interface
C5.0	C5.0 Release 2.06 algorithm with rule set option in SPSS
CAL. BUS & PROF. CODE	California Business and Professions Code
CAL. CIV. CODE	California Civil Code
CASIA-HWDB	Institute of Automation of the Chinese Academy of Sciences-Handwriting Databases
CAR	Computer-assisted reporting
CART	Classification and regression trees algorithm in MATLAB's classregtree function
CDBN	Convolutional deep belief network
CNN	Convolutional neural network
CONN. GEN. STAT. ANN.	Connecticut general statutes annotated
CS	Compressed sensing
CVPR***	Computer vision and pattern recognition
DAS	Domain awareness system
DBN	Deep belief network
DCNN	Deep convolutional neural network
DHS	US Department of Homeland Security
DList	Decision lists algorithm in SPSS
DNA	Deoxyribonucleic acid
DNNs	Deep neural networks
DTD	Describable Textures Dataset

EU	European Union
EU GSPR	European Union General Data Protection Regulation
EDPS	European Data Protection Supervisor
EFF	Electronic Frontier Foundation
EUCJ	European Union Court of Justice
EULA	End user license agreement
FLIC	Frames Labeled in Cinema
FMD	Flickr Material Database
FTC	Federal Trade Commission
GA. CODE ANN.	Code of Georgia Annotated
GDPR	General Data Protection Regulation
GPS	Global Positioning System
GSM	Global System for Mobile Communications
GSMA	GSM Association
GTSRB	German Traffic Sign Recognition Benchmark
HCI	Human-computer interaction
HDI	Human-data interaction
ICCV***	International Conference on Computer Vision
ICT	Information and communications technology
IEEE***	Institute of Electrical and Electronics Engineers
ILPD	Indian Liver Patient Dataset
Ionos	The Ionosphere Dataset
IP	Integer programming
IRB	Institutional review board
ISLVRC	ImageNet Large-Scale Visual Recognition Challenge
kNN	The k-nearest neighbor algorithm in SPSS
Liver	BUPA Liver Disorders Dataset
LFW	Labeled Faces in the Wild
LP	Linear programming
LSP	Leeds Sports Pose
MCDNN	Multicolumn deep neural network
MNIST	Mixed National Institute of Standards and Technology
NP-hard	Non-deterministic Polynomial-time
NSA	National Security Agency
NHS	National Health Service
NIPS***	Neural information processing systems
NPR	National Public Radio
Parkin	Parkinson's Dataset
PETs	Privacy-enhancing technologies
PGP	Pretty Good Privacy
Pima	Pima Indian Diabetes Dataset
PPTCs	Privacy policy terms and conditions

QII	Quantitative input influence
RTDNA	Radio Television Digital News Association
RuB	Boosting approach rule learner
RuSC	Set covering approach rule learner
SCM	Set covering machine
SDNY	United States District Court for the Southern District of New York
Sonar	Connectionist bench sonar dataset
SQGT	Semiquantitative group testing
SRF	Schweizer Radio und Fernsehen
SVM	Support vector machine
SUS	Secondary Uses Service
T3	Tastes, Ties, and Time
t-SNE	Stochastic neighbor embedding
TGT	Threshold group testing
ToS	Terms of service
Trans	Blood Transfusion Service Center Dataset
TrBag	The random forests classifier in MATLAB's Tree-Bagger class
UCI	University of California, Irvine
VOC	Visual object classes
WAF	We Are Family
WDBC	Wisconsin Diagnostic Breast Cancer Dataset
WEF	World Economic Forum
WPF	World Privacy Forum
YTF	YouTube Faces

Note that acronyms marked with *** are never used without the long form in the text.

The original version of this book was revised. An erratum to this book can be found at http://dx.doi.org/10.1007/978-3-319-54024-5_10

Part I
Transparent Mining

The Tyranny of Data? The Bright and Dark Sides of Data-Driven Decision-Making for Social Good

Bruno Lepri, Jacopo Staiano, David Sangokoya, Emmanuel Letouzé, and Nuria Oliver

Abstract The unprecedented availability of large-scale human behavioral data is profoundly changing the world we live in. Researchers, companies, governments, financial institutions, non-governmental organizations and also citizen groups are actively experimenting, innovating and adapting algorithmic decision-making tools to understand global patterns of human behavior and provide decision support to tackle problems of societal importance. In this chapter, we focus our attention on *social good decision-making algorithms*, that is algorithms strongly influencing decision-making and resource optimization of public goods, such as public health, safety, access to finance and fair employment. Through an analysis of specific use cases and approaches, we highlight both the positive opportunities that are created through data-driven algorithmic decision-making, and the potential negative consequences that practitioners should be aware of and address in order to truly realize the potential of this emergent field. We elaborate on the need for these algorithms to provide transparency and accountability, preserve privacy and be tested and evaluated *in context*, by means of living lab approaches involving citizens. Finally, we turn to the requirements which would make it possible to leverage the predictive power of data-driven human behavior analysis while ensuring transparency, accountability, and civic participation.

B. Lepri (✉)
Fondazione Bruno Kessler, Trento, Italy
e-mail: lepri@fbk.eu

J. Staiano
Fortia Financial Solutions, Paris, France
e-mail: jacopo.staiano@fortia.fr

D. Sangokoya • N. Oliver
Data-Pop Alliance, New York, NY, USA
e-mail: dsangokoya@datapopalliance.org; nuria@alum.mit.edu

E. Letouzé
Data-Pop Alliance, New York, NY, USA

MIT Media Lab, Cambridge, MA, USA
e-mail: eletouze@mit.edu

T. Cerquitelli et al. (eds.), *Transparent Data Mining for Big and Small Data*,
Studies in Big Data 32, DOI 10.1007/978-3-319-54024-5_1

1 Introduction

The world is experiencing an unprecedented transition where human behavioral data has evolved from being a scarce resource to being a massive and real-time stream. This availability of large-scale data is profoundly changing the world we live in and has led to the emergence of a new discipline called *computational social science* [45]; finance, economics, marketing, public health, medicine, biology, politics, urban science and journalism, to name a few, have all been disrupted to some degree by this trend [41].

Moreover, the automated analysis of anonymized and aggregated large-scale human behavioral data offers new possibilities to understand global patterns of human behavior and to help decision makers tackle problems of societal importance [45], such as monitoring socio-economic deprivation [8, 75, 76, 88] and crime [10, 11, 84, 85, 91], mapping the propagation of diseases [37, 94], or understanding the impact of natural disasters [55, 62, 97]. Thus, researchers, companies, governments, financial institutions, non-governmental organizations and also citizen groups are actively experimenting, innovating and adapting algorithmic decision-making tools, often relying on the analysis of personal information.

However, researchers from different disciplinary backgrounds have identified a range of social, ethical and legal issues surrounding data-driven decision-making, including privacy and security [19, 22, 24, 56], transparency and accountability [18, 61, 99, 100], and bias and discrimination [3, 79]. For example, Barocas and Selbst [3] point out that the use of data-driven decision making processes can result in disproportionate adverse outcomes for disadvantaged groups, in ways that look like discrimination. Algorithmic decisions can reproduce patterns of discrimination, due to decision makers' prejudices [60], or reflect the biases present in the society [60]. In 2014, the White House released a report, titled "Big Data: Seizing opportunities, preserving values" [65] that highlights the discriminatory potential of big data, including how it could undermine longstanding civil rights protections governing the use of personal information for credit, health, safety, employment, etc. For example, data-driven decisions about applicants for jobs, schools or credit may be affected by hidden biases that tend to flag individuals from particular demographic groups as unfavorable for such opportunities. Such outcomes can be self-reinforcing, since systematically reducing individuals' access to credit, employment and educational opportunities may worsen their situation, which can play against them in future applications.

In this chapter, we focus our attention on *social good algorithms*, that is algorithms strongly influencing decision-making and resource optimization of public goods, such as public health, safety, access to finance and fair employment. These algorithms are of particular interest given the magnitude of their impact on quality of life and the risks associated with the information asymmetry surrounding their governance.

In a recent book, William Easterly evaluates how global economic development and poverty alleviation projects have been governed by a "tyranny of experts"—in this case, aid agencies, economists, think tanks and other analysts—who consistently favor top-down, technocratic governance approaches at the expense of the individual rights of citizens [28]. Easterly details how these experts reduce multidimensional social phenomena such as poverty or justice into a set of technical solutions that do not take into account either the political systems in which they operate or the rights of intended beneficiaries. Take for example the displacement of farmers in the Mubende district of Uganda: as a direct result of a World Bank project intended to raise the region's income by converting land to higher value uses, farmers in this district were forcibly removed from their homes by government soldiers in order to prepare for a British company to plant trees in the area [28]. Easterly underlines the cyclic nature of this tyranny: technocratic justifications for specific interventions are considered objective; intended beneficiaries are unaware of the opaque, black box decision-making involved in these resource optimization interventions; and experts (and the coercive powers which employ them) act with impunity and without redress.

If we turn to the use, governance and deployment of big data approaches in the public sector, we can draw several parallels towards what we refer to as the "tyranny of data", that is the adoption of data-driven decision-making under the technocratic and top-down approaches highlighted by Easterly [28]. We elaborate on the need for *social good decision-making algorithms* to provide transparency and accountability, to only use personal information—owned and controlled by individuals—with explicit consent, to ensure that privacy is preserved when data is analyzed in aggregated and anonymized form, and to be tested and evaluated *in context*, that is by means of living lab approaches involving citizens. In our view, these characteristics are crucial for fair data-driven decision-making as well as for citizen engagement and participation.

In the rest of this chapter, we provide the readers with a compendium of the issues arising from current big data approaches, with a particular focus on specific use cases that have been carried out to date, including urban crime prediction [11], inferring socioeconomic status of countries and individuals [8, 49, 76], mapping the propagation of diseases [37, 94] and modeling individuals' mental health [9, 20, 47]. Furthermore, we highlight factors of risk (e.g. privacy violations, lack of transparency and discrimination) that might arise when decisions potentially impacting the daily lives of people are heavily rooted in the outcomes of black-box data-driven predictive models. Finally, we turn to the requirements which would make it possible to leverage the predictive power of data-driven human behavior analysis while ensuring transparency, accountability, and civic participation.

2 The Rise of Data-Driven Decision-Making for Social Good

The unprecedented stream of large-scale, human behavioral data has been described as a "tidal wave" of opportunities to both predict and act upon the analysis of the petabytes of digital signals and traces of human actions and interactions. With such massive streams of relevant data to mine and train algorithms with, as well as increased analytical and technical capacities, it is of no surprise that companies and public sector actors are turning to machine learning-based algorithms to tackle complex problems at the limits of human decision-making [36, 96]. The history of human decision-making—particularly when it comes to questions of power in resource allocation, fairness, justice, and other public goods—is wrought with innumerable examples of extreme bias, leading towards corrupt, inefficient or unjust processes and outcomes [2, 34, 70, 87]. In short, human decision-making has shown significant limitations and the turn towards data-driven algorithms reflects a search for objectivity, evidence-based decision-making, and a better understanding of our resources and behaviors.

Diakopoulos [27] characterizes the function and power of algorithms in four broad categories: (1) *classification*, the categorization of information into separate "classes", based on its features; (2) *prioritization*, the denotation of emphasis and rank on particular information or results at the expense of others based on a pre-defined set of criteria; (3) *association*, the determination of correlated relationships between entities; and (4) *filtering*, the inclusion or exclusion of information based on pre-determined criteria.

Table 1 provides examples of types of algorithms across these categories.

This chapter places emphasis on what we call social good algorithms—algorithms strongly influencing decision-making and resource optimization for public goods. These algorithms are designed to analyze massive amounts of human behavioral data from various sources and then, based on pre-determined criteria, select the information most relevant to their intended purpose. While resource allocation and decision optimization over limited resources remain common features of the public sector, the use of social good algorithms brings to a new level the amount of human behavioral data that public sector actors can access, the capacities with which they can analyze this information and deliver results, and the communities of experts and common people who hold these results to be objective. The ability of these algorithms to identify, select and determine

Table 1 Algorithmic function and examples, adapted from Diakopoulos [27] and Latzer et al. [44]

Function	Type	Examples
Prioritization	General and search engines, meta search engines, semantic search engines, questions & answers services	Google, Bing, Baidu; image search; social media; Quora; Ask.com
Classification	Reputation systems, news scoring, credit scoring, social scoring	Ebay, Uber, Airbnb; Reddit, Digg; CreditKarma; Klout
Association	Predicting developments and trends	ScoreAhit, Music Xray, Google Flu Trends
Filtering	Spam filters, child protection filters, recommender systems, news aggregators	Norton; Net Nanny; Spotify, Netflix; Facebook Newsfeed

information of relevance beyond the scope of human decision-making creates a new kind of decision optimization facilitated by both the design of the algorithms and the data from which they are based. However, as discussed later in the chapter, this new process is often opaque and assumes a level of impartiality that is not always accurate. It also creates information asymmetry and lack of transparency between actors using these algorithms and the intended beneficiaries whose data is being used.

In the following sections, we assess the nature, function and impact of the use of social good algorithms in three key areas: criminal behavior dynamics and predictive policing; socio-economic deprivation and financial inclusion; and public health.

2.1 Criminal Behavior Dynamics and Predictive Policing

Researchers have turned their attention to the automatic analysis of criminal behavior dynamics both from *people-* and *place-*centric perspectives. The people-centric perspective has mostly been used for individual or collective criminal profiling [67, 72, 90]. For example, Wang et al. [90] proposed a machine learning approach, called Series Finder, to the problem of detecting specific patterns in crimes that are committed by the same offender or group of offenders.

In 2008, the criminologist David Weisburd proposed a shift from a people-centric paradigm of police practices to a place-centric one [93], thus focusing on geographical topology and micro-structures rather than on criminal profiling. An example of a place-centric perspective is the detection, analysis, and interpretation of crime hotspots [16, 29, 53]. Along these lines, a novel application of quantitative tools from mathematics, physics and signal processing has been proposed by Toole et al. [84] to analyse spatial and temporal patterns in criminal offense records. Their analyses of crime data from 1991 to 1999 for the American city of Philadelphia indicated the existence of multi-scale complex relationships in space and time. Further, over the last few years, aggregated and anonymized mobile phone data has opened new possibilities to study city dynamics with unprecedented temporal and spatial granularities [7]. Recent work has used this type of data to predict crime hotspots through machine-learning algorithms [10, 11, 85].

More recently, these *predictive policing* approaches [64] are moving from the academic realm (universities and research centers) to police departments. In Chicago, police officers are paying particular attention to those individuals flagged, through risk analysis techniques, as most likely to be involved in future violence. In Santa Cruz, California, the police have reported a dramatic reduction in burglaries after adopting algorithms that predict where new burglaries are likely to occur. In Charlotte, North Carolina, the police department has generated a map of high-risk areas that are likely to be hit by crime. The Police Departments of Los Angeles, Atlanta and more than 50 other cities in the US are using PredPol, an algorithm that generates 500 by 500 square foot predictive boxes on maps, indicating areas where crime is most likely to occur. Similar approaches have also been implemented in Brasil, the UK and the Netherlands. Overall, four main predictive policing approaches are currently being used: (1) methods to forecast places and times with an increased risk of crime [32], (2) methods to detect offenders and flag individuals at risk of offending in the future [64], (3) methods to identify perpetrators [64], and (4) methods to identify groups or, in some cases, individuals who are likely to become the victims of crime [64].

2.2 Socio-Economic Deprivation and Financial Inclusion

Being able to accurately measure and monitor key sociodemographic and economic indicators is critical to design and implement public policies [68]. For example, the geographic distribution of poverty and wealth is used by governments to make decisions about how to allocate scarce resources and provides a foundation for the study of the determinants of economic growth [33, 43]. The quantity and quality of economic data available have significantly improved in recent years. However, the scarcity of reliable key measures in developing countries represents a major challenge to researchers and policy-makers,[1] thus hampering efforts to target interventions effectively to areas of greatest need (e.g. African countries) [26, 40]. Recently, several researchers have started to use mobile phone data [8, 49, 76], social media [88] and satellite imagery [39] to infer the poverty and wealth of individual subscribers, as well as to create high-resolution maps of the geographic distribution of wealth and deprivation.

The use of novel sources of behavioral data and algorithmic decision-making processes is also playing a growing role in the area of financial services, for example credit scoring. Credit scoring is a widely used tool in the financial sector to compute the risks of lending to potential credit customers. Providing information about the ability of customers to pay back their debts or conversely to default, credit scores have become a key variable to build financial models of customers. Thus, as lenders have moved from traditional interview-based decisions to data-driven models to assess credit risk, consumer lending and credit scoring have become increasingly sophisticated. Automated credit scoring has become a standard input into the pricing of mortgages, auto loans, and unsecured credit. However, this approach is mainly based on the past financial history of customers (people or businesses) [81], and thus not adequate to provide credit access to people or businesses when no financial history is available. Therefore, researchers and companies are investigating novel sources of data to replace or to improve traditional credit scores, potentially opening credit access to individuals or businesses that traditionally have had poor or no access to mainstream financial services—e.g. people who are unbanked or underbanked, new immigrants, graduating students, etc. Researchers have leveraged mobility patterns from credit card transactions [74]

[1] http://www.undatarevolution.org/report/.

and mobility and communication patterns from mobile phones to automatically build user models of spending behavior [73] and propensity to credit defaults [71, 74]. The use of mobile phone, social media, and browsing data for financial risk assessment has also attracted the attention of several entrepreneurial efforts, such as Cignifi,[2] Lenddo,[3] InVenture,[4] and ZestFinance.[5]

2.3 Public Health

The characterization of individuals and entire populations' mobility is of paramount importance for public health [57]: for example, it is key to predict the spatial and temporal risk of diseases [35, 82, 94], to quantify exposure to air pollution [48], to understand human migrations after natural disasters or emergency situations [4, 50], etc. The traditional approach has been based on household surveys and information provided from census data. These methods suffer from recall bias and limitations in the size of the population sample, mainly due to excessive costs in the acquisition of the data. Moreover, survey or census data provide a snapshot of the population dynamics at a given moment in time. However, it is fundamental to monitor mobility patterns in a continuous manner, in particular during emergencies in order to support decision making or assess the impact of government measures.

Tizzoni et al. [82] and Wesolowski et al. [95] have compared traditional mobility surveys with the information provided by mobile phone data (Call Detail Records or CDRs), specifically to model the spread of diseases. The findings of these works recommend the use of mobile phone data, by themselves or in combination with traditional sources, in particular in low-income economies where the availability of surveys is highly limited.

Another important area of opportunity within public health is mental health. Mental health problems are recognized to be a major public health issue.[6] "However, the traditional model of episodic care is suboptimal to prevent mental health outcomes and improve chronic disease outcomes. In order to assess human behavior in the context of mental wellbeing, the standard clinical practice relies on periodic self-reports" that suffer from subjectivity and memory biases, and are likely influenced by the current mood state. Moreover, individuals with mental conditions typically visit doctors when the crisis has already happened and thus report limited information about precursors useful to prevent the crisis onset. These novel sources of behavioral data yield the possibility of monitoring mental health-related behaviors and symptoms outside of clinical settings and without having to

[2] http://cignifi.com/.

[3] https://www.lenddo.com/.

[4] http://tala.co/.

[5] https://www.zestfinance.com/.

[6] http://www.who.int/topics/mental_health/en/.

Table 2 Summary table for the literature discussed in Sect. 2

Key area	Problems tackled	References
Predictive policing	Criminal behavior profiling	Ratcliffe [67], Short et al. [72], Wang et al. [90]
	Crime hotspot prediction	Bogomolov et al. [10, 11], Ferguson [32], Traunmueller et al. [85]
	Perpetrator(s)/victim(s) identification	Perry et al. [64]
Finance & economy	Wealth & deprivation mapping	Blumenstock et al. [8], Louail [49], Jean et al. [39], Soto et al. [76], Venerandi et al. [88]
	Spending behavior profiling	Singh et al. [73]
	Credit scoring	San Pedro et al. [71], Singh et al. [74]
Public health	Epidemiologic studies	Frias-Martinez et al. [35], Tizzoni et al. [82], Wesolowski et al. [94]
	Environment and emergency mapping	Bengtsson et al. [4], Liu et al. [48], Lu et al. [50]
	Mental health	Bogomolov et al. [9], De Choudhury [20], de Oliveira et al. [25], Faurholt-Jepsena et al. [30], Lepri et al. [46], LiKamWa et al. [47], Matic and Oliver [52], Osmani et al. [59]

depend on self-reported information [52]. For example, several studies have shown that behavioral data collected through mobile phones and social media can be exploited to recognize bipolar disorders [20, 30, 59], mood [47], personality [25, 46] and stress [9].

Table 2 summarizes the main points emerging from the literature reviewed in this section.

3 The Dark Side of Data-Driven Decision-Making for Social Good

The potential positive impact of big data and machine learning-based approaches to decision-making is huge. However, several researchers and experts [3, 19, 61, 79, 86] have underlined what we refer to as *the dark side* of data-driven decision-making, including violations of privacy, information asymmetry, lack of transparency, discrimination and social exclusion. In this section we turn our attention to these elements before outlining three key requirements that would be necessary in order to realize the positive impact, while minimizing the potential negative consequences of data-driven decision-making in the context of social good.

3.1 Computational Violations of Privacy

Reports and studies [66] have focused on the misuse of personal data disclosed by users and on the aggregation of data from different sources by entities playing as data brokers with direct implications in privacy. An often overlooked element is that the computational developments coupled with the availability of novel sources of behavioral data (e.g. social media data, mobile phone data, etc.) now allow inferences about private information that may never have been disclosed. This element is essential to understand the issues raised by these algorithmic approaches.

A recent study by Kosinski et al. [42] combined data on Facebook "Likes" and limited survey information to accurately predict a male user's sexual orientation, ethnic origin, religious and political preferences, as well as alcohol, drug, and cigarette use. Moreover, Twitter data has recently been used to identify people with a high likelihood of falling into depression before the onset of the clinical symptoms [20].

It has also been shown that, despite the algorithmic advancements in anonymizing data, it is feasible to infer identities from anonymized human behavioral data, particularly when combined with information derived from additional sources. For example, Zang et al. [98] have reported that if home and work addresses were available for some users, up to 35% of users of the mobile network could be deanonymized just using the two most visited towers, likely to be related to their home and work location. More recently, de Montjoye et al. [22, 24] have demonstrated how unique mobility and shopping behaviors are for each individual. Specifically, they have shown that four spatio-temporal points are enough to uniquely identify 95% of people in a mobile phone database of 1.5M people and to identify 90% of people in a credit card database of 1M people.

3.2 Information Asymmetry and Lack of Transparency

Both governments and companies use data-driven algorithms for decision making and optimization. Thus, accountability in government and corporate use of such decision-making tools is fundamental in both validating their utility toward the public interest as well as redressing harms generated by these algorithms.

However, the ability to accumulate and manipulate behavioral data about customers and citizens on an unprecedented scale may give big companies and intrusive/authoritarian governments powerful means to manipulate segments of the population through targeted marketing efforts and social control strategies. In particular, we might witness an *information asymmetry* situation where a powerful few have access and use knowledge that the majority do not have access to, thus leading to an—or exacerbating the existing—asymmetry of power between the state or big companies on one side and the people on the other side [1]. In addition, the nature and the use of various data-driven algorithms for social good, as well as the lack of computational or data literacy among citizens, makes algorithmic transparency difficult to generalize and accountability difficult to assess [61].

Burrell [12] has provided a useful framework to characterize three different types of opacity in algorithmic decision-making: (1) *intentional opacity*, whose objective is the protection of the intellectual property of the inventors of the algorithms. This type of opacity could be mitigated with legislation that would force decision-makers towards the use of open source systems. The new General Data Protection Regulation (GDPR) in the EU with a "right to an explanation" starting in 2018 is an example of such legislation.[7] However, there are clear corporate and governmental interests in favor of intentional opacity which make it difficult to eliminate this type of opacity; (2) *illiterate opacity*, due to the fact that the vast majority of people lack the technical skills to understand the underpinnings of algorithms and machine learning models built from data. This kind of opacity might be attenuated with stronger education programs in computational thinking and by enabling that independent experts advise those affected by algorithm decision-making; and (3) *intrinsic opacity*, which arises by the nature of certain machine learning methods that are difficult to interpret (e.g. deep learning models). This opacity is well known in the machine learning community (usually referred to as the *interpretability problem*). The main approach to combat this type of opacity requires using alternative machine learning models that are easy to interpret by humans, despite the fact that they might yield lower accuracy than black-box non-interpretable models.

Fortunately, there is increasing awareness of the importance of reducing or eliminating the opacity of data-driven algorithmic decision-making systems. There are a number of research efforts and initiatives in this direction, including the Data Transparency Lab[8] which is a "community of technologists, researchers, policymakers and industry representatives working to advance online personal data transparency through research and design", and the DARPA Explainable Artificial Intelligence (XAI) project.[9] A tutorial on the subject has been held at the 2016 ACM Knowledge and Data Discovery conference [38]. Researchers from New York University's Information Law Institute, such as Helen Nissenbaum and Solon Barocas, and Microsoft Research, such as Kate Crawford and Tarleton Gillespie, have held several workshops and conferences during the past few years on the ethical and legal challenges related to algorithmic governance and decision-making.[10] A nominee for the National Book Award, Cathy O'Neil's book, "Weapons of Math Destruction," details several case studies on harms and risks to public accountability associated with big data-driven algorithmic decision-making, particularly in the areas of criminal justice and education [58]. Recently, in partnership with Microsoft

[7]Regulation (EU) 2016/679 of the European Parliament and of the Council of 27 April 2016 on the protection of natural persons with regard to the processing of personal data and on the free movement of such data, and repealing Directive 95/46/EC (General Data Protection Regulation) http://eur-lex.europa.eu/eli/reg/2016/679/oj.

[8]http://www.datatransparencylab.org/.

[9]http://www.darpa.mil/program/explainable-artificial-intelligence.

[10]http://www.law.nyu.edu/centers/ili/algorithmsconference.

Research and others, the White House Office of Science and Technology Policy has co-hosted several public symposiums on the impacts and challenges of algorithms and artificial intelligence, specifically in social inequality, labor, healthcare and ethics.[11]

3.3 *Social Exclusion and Discrimination*

From a legal perspective, Tobler [83] argued that discrimination derives from "the application of different rules or practices to comparable situations, or of the same rule or practice to different situations". In a recent paper, Barocas and Selbst [3] elaborate that discrimination may be an artifact of the data collection and analysis process itself; more specifically, even with the best intentions, data-driven algorithmic decision-making can lead to discriminatory practices and outcomes. Algorithmic decision procedures can reproduce existing patterns of discrimination, inherit the prejudice of prior decision makers, or simply reflect the widespread biases that persist in society [19]. It can even have the perverse result of exacerbating existing inequalities by suggesting that historically disadvantaged groups actually deserve less favorable treatment [58].

Discrimination from algorithms can occur for several reasons. First, input data into algorithmic decisions may be poorly weighted, leading to *disparate impact*; for example, as a form of *indirect discrimination*, overemphasis of zip code within predictive policing algorithms can lead to the association of low-income African-American neighborhoods with areas of crime and as a result, the application of specific targeting based on group membership [17]. Second, discrimination can occur from the decision to use an algorithm itself. Categorization—through algorithmic classification, prioritization, association and filtering—can be considered as a form of *direct discrimination*, whereby algorithms are used for disparate treatment [27]. Third, algorithms can lead to discrimination as a result of the misuse of certain models in different contexts [14]. Fourth, in a form of feedback loop, biased training data can be used both as evidence for the use of algorithms and as proof of their effectiveness [14].

The use of algorithmic data-driven decision processes may also result in individuals mistakenly being denied opportunities based not on their own action but on the actions of others with whom they share some characteristics. For example, some credit card companies have lowered a customer's credit limit, not based on the customer's payment history, but rather based on analysis of other customers with a poor repayment history that had shopped at the same establishments where the customer had shopped [66].

Indeed, we find increasing evidence of detrimental impact already taking place in current non-algorithmic approaches to credit scoring and generally to background

[11] https://www.whitehouse.gov/blog/2016/05/03/preparing-future-artificial-intelligence.

checks. The latter have been widely used in recent years in several contexts: it is common to agree to be subjected to background checks when applying for a job, to lease a new apartment, etc. In fact, hundreds of thousands of people have unknowingly seen themselves adversely affected on existential matters such as job opportunities and housing availability due to simple but common mistakes (for instance, misidentification) in the procedures used by external companies to perform background checks.[12] It is worth noticing that the trivial procedural mistakes causing such adverse outcomes are bound to disappear once fully replaced with data-driven methodologies. Alas, this also means that should such methodologies not be transparent in their inner workings, the effects are likely to stay though with different roots. Further, the effort required to identify the causes of unfair and discriminative outcomes can be expected to be exponentially larger, as exponentially more complex will be the black-box models employed to assist in the decision-making process. This scenario highlights particularly well the need for machine learning models featuring transparency and accountability: adopting black-box approaches in scenarios where the lives of people would be seriously affected by a machine-driven decision could lead to forms of *algorithmic stigma*,[13] a particularly creepy scenario considering that those stigmatized might never become aware of being so, and the stigmatizer will be an unaccountable machine. Recent advances in neural network-based (deep learning) models are yielding unprecedented accuracies in a variety of fields. However, such models tend to be difficult—if not impossible—to interpret, as previously explained. In this chapter, we highlight the need for data-driven machine learning models that are interpretable by humans when such models are going to be used to make decisions that affect individuals or groups of individuals.

4 Requirements for Positive Disruption of Data-Driven Policies

As noted in the previous sections, both governments and companies are increasingly using data-driven algorithms for decision support and resource optimization. In the context of social good, accountability in the use of such powerful decision support tools is fundamental in both validating their utility toward the public interest as well as redressing corrupt or unjust harms generated by these algorithms. Several scholars have emphasized elements of what we refer to as the dark side of data-

[12] See, for instance, http://www.chicagotribune.com/business/ct-background-check-penalties-1030-biz-20151029-story.html.

[13] As a social phenomenon, the concept of stigma has received significant attention by sociologists, who under different frames highlighted and categorized the various factors leading individuals or groups to be discriminated against by society, the countermoves often adopted by the stigmatized, and analyzed dynamics of reactions and evolution of stigma. We refer the interested reader to the review provided by Major and O'Brian [51].

driven policies for social good, including violations of individual and group privacy, information asymmetry, lack of transparency, social exclusion and discrimination. Arguments against the use of social good algorithms typically call into question the use of machines in decision support and the need to protect the role of human decision-making.

However, therein lies a huge potential and imperative for leveraging large scale human behavioral data to design and implement policies that would help improve the lives of millions of people. Recent debates have focused on characterizing data-driven policies as either "good" or "bad" for society. We focus instead on the potential of data-driven policies to lead to positive disruption, such that they reinforce and enable the powerful functions of algorithms as tools generating value while minimizing their dark side.

In this section, we present key *human-centric* requirements for positive disruption, including a fundamental renegotiation of user-centric data ownership and management, the development of tools and participatory infrastructures towards increased algorithmic transparency and accountability, and the creation of living labs for experimenting and co-creating data-driven policies. We place humans at the center of our discussion as humans are ultimately both the actors and the subjects of the decisions made via algorithmic means. If we are able to ensure that these requirements are met, we should be able to realize the positive potential of data-driven algorithmic decision-making while minimizing the risks and possible negative unintended consequences.

4.1 User-Centric Data Ownership and Management

A big question on the table for policy-makers, researchers, and intellectuals is: *how do we unlock the value of human behavioral data while preserving the fundamental right to privacy?* This question implicitly recognizes the risks, in terms not only of possible abuses but also of a "missed chance for innovation", inherent to the current paradigm: the dominant *siloed* approach to data collection, management, and exploitation, precludes participation to a wide range of actors, most notably to the very producers of personal data (i.e. the users).

On this matter, new user-centric models for personal data management have been proposed, in order to empower individuals with more control of their own data's life-cycle [63]. To this end, researchers and companies are developing repositories which implement medium-grained access control to different kinds of personally identifiable information (PII), such as passwords, social security numbers and health data [92], location [23] and personal data collected by means of smartphones or connected devices [23]. A pillar of these approaches is represented by a *Personal Data Eco-system*, composed by secure vaults of personal data whose owners are granted full control of.

Along this line, an interesting example is the Enigma platform [101] that leverages the recent technological trend of decentralization: advances in the fields of cryptography and decentralized computer networks have resulted in the emergence of a novel technology—known as the *blockchain*—which has the potential to reduce the role of one of the most important actors in our society: the middle man [5, 21]. By allowing people to transfer a unique piece of digital property or data to others, in a safe, secure, and immutable way, this technology can create digital currencies (e.g. bitcoin) that are not backed by any governmental body [54]; self-enforcing digital contracts, called smart contracts, whose execution does not require any human intervention (e.g. Ethereum) [80]; and decentralized marketplaces that aim to operate free from regulations [21]. Hence, Enigma tackles the challenge of providing a secure and trustworthy mechanism for the exchange of goods in a personal data market. To illustrate how the platform works, consider the following example: a group of data analysts of an insurance company wishes to test a model that leverages people's mobile phone data. Instead of sharing their raw data with the data analysts in the insurance company, the users can securely store their data in Enigma, and only provide the data analysts with a permission to execute their study. The data analysts are thus able to execute their code and obtain the results, but nothing else. In the process, the users are compensated for having given access to their data and the computers in the network are paid for their computing resources [78].

4.2 Algorithmic Transparency and Accountability

The deployment of a machine learning model entails a degree of *trust* on how satisfactory its performance in the wild will be from the perspectives of both the builders and the users. Such trust is assessed at several points during an iterative model building process. Nonetheless, many of the state-of-the-art machine learning-based models (e.g. neural networks) act as black-boxes once deployed. When such models are used for decision-making, the lack of explanations regarding why and how they have reached their decisions poses several concerns. In order to address this limitation, recent research efforts in the machine learning community have proposed different approaches to make the algorithms more amenable to *ex ante* and *ex post* inspection. For example, a number of studies have attempted to tackle the issue of discrimination within algorithms by introducing tools to both identify [6] and rectify [6, 13, 31] cases of unwanted bias. Recently, Ribeiro et al. [69] have proposed a model-agnostic method to derive explanations for the predictions of a given model.

An interesting ongoing initiative is the Open Algorithms (OPAL) project,[14] a multi-partner effort led by Orange, the MIT Media Lab, Data-Pop Alliance, Imperial College London, and the World Economic Forum, that aims to open—

[14]http://datapopalliance.org/open-algorithms-a-new-paradigm-for-using-private-data-for-social-good/.

without exposing—data collected and stored by private companies by "sending the code to the data" rather than the other way around. The goal is to enable the design, implementation and monitoring of development policies and programs, accountability of government action, and citizen engagement while leveraging the availability of large scale human behavioral data. OPAL's core will consist of an open platform allowing open algorithms to run on the servers of partner companies, behind their firewalls, to extract key development indicators and operational data of relevance for a wide range of potential users. Requests for approved, certified and pre-determined indicators by third parties—e.g. mobility matrices, poverty maps, population densities—will be sent to them via the platform; certified algorithms will run on the data in a multiple privacy-preserving manner, and results will be made available via an API. The platform will also be used to foster civic engagement of a broad range of social constituents—academic institutions, private sector companies, official institutions, non-governmental and civil society organizations. Overall, the OPAL initiative has three key objectives: (1) engage with data providers, users, and analysts at all the stages of algorithm development; (2) contribute to building local capacities and help shaping the future technological, ethical and legal frameworks that will govern the collection, control and use of human behavioral data to foster social progress; and (3) build data literacy among users and partners, conceptualized as "the ability to constructively engage in society through and about data". Initiatives such as OPAL have the potential to enable more human-centric accountable and transparent data-driven decision-making and governance.

4.3 Living Labs to Experiment with Data-Driven Policies

The use of real-time human behavioral data to design and implement policies has been traditionally outside the scope of policymaking work. However, the potential of this type of data will only be realized when policy makers are able to analyze the data, to study human behavior and to test policies in the real world. A possible way is to build living laboratories—communities of volunteers willing to try new ways of doing things in a natural setting—in order to test ideas and hypotheses in a real life setting. An example is the Mobile Territorial Lab (MTL), a living lab launched by Fondazione Bruno Kessler, Telecom Italia, the MIT Media Lab and Telefonica, that has been observing the lives of more than 100 families through multiple channels for more than 3 years [15]. Data from multiple sources, including smartphones, questionnaires, experience sampling probes, etc. has been collected and used to create a multi-layered view of the lives of the study participants. In particular, social interactions (e.g. call and SMS communications), mobility routines and spending patterns, etc. have been captured. One of the MTL goals is to devise new ways of sharing personal data by means of Personal Data Store (PDS) technologies, in order to promote greater civic engagement. An example of an application enabled by PDS technologies is the sharing of best practices among families with young children. How do other families spend their money? How much do they get out and socialize? Once the individual gives permission, MyDataStore [89], the PDS system used by MTL participants, allows such personal data to be collected, anonymized, and shared with other young families safely and automatically.

The MTL has been also used to investigate how to deal with the sensitivities of collecting and using deeply personal data in real-world situations. In particular, an MTL study investigated the perceived monetary value of mobile information and its association with behavioral characteristics and demographics; the results corroborate the arguments towards giving back to the people (users, citizens, according to the scenario) control on the data they constantly produce [77].

Along these lines, Data-Pop Alliance and the MIT Media Lab launched in May 2016 a novel initiative called "Laboratorio Urbano" in Bogotá, Colombia, in partnership with Bogotá's city government and Chamber of Commerce. The main objective of the Bogotá Urban Laboratory is to contribute to the city's urban vitality, with a focus on mobility and safety, through collaborative research projects and dialogues involving the public and private sectors, academic institutions, and citizens. Similar initiatives are being planned in other major cities of the global south, including Dakar, Senegal, with the goal of strengthening and connecting local ecosystems where data-driven innovations can take place and scale.

Figure 1 provides the readers with a visual representation of the factors playing a significant role in positive data-driven disruption.

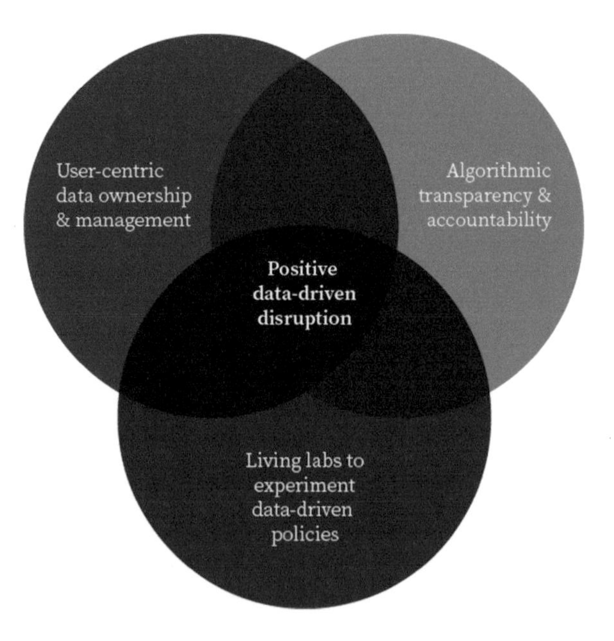

Fig. 1 Requirements summary for positive data-driven disruption

5 Conclusion

In this chapter we have provided an overview of both the opportunities and the risks of data-driven algorithmic decision-making for the public good. We are witnessing an unprecedented time in our history, where vast amounts of fine grained human behavioral data are available. The analysis of this data has the potential to help inform policies in public health, disaster management, safety, economic development and national statistics among others. In fact, the use of data is at the core of the 17 Sustainable Development Goals (SDGs) defined by United Nations, both in order to achieve the goals and to measure progress towards their achievement.

While this is an exciting time for researchers and practitioners in this new field of computational social sciences, we need to be aware of the risks associated with these new approaches to decision making, including violation of privacy, lack of transparency, information asymmetry, social exclusion and discrimination. We have proposed three human-centric requirements that we consider to be of paramount importance in order to enable positive disruption of data-driven policy-making: user-centric data ownership and management; algorithmic transparency and accountability; and living labs to experiment with data-driven policies in the wild. It will be only when we honor these requirements that we will be able to move from the feared tyranny of data and algorithms to a data-enabled model of democratic governance running against tyrants and autocrats, and for the people.

References

1. Akerlof, G.A.: The market for "lemons": quality uncertainty and the market mechanism. Q. J. Econ. **84**(3), 488–500 (1970)
2. Akerlof, G.A., Shiller, R.J.: Animal Spirits: How Human Psychology Drives the Economy, and Why It Matters for Global Capitalism. Princeton University Press, Princeton (2009)
3. Barocas, S., Selbst, A.D.: Big data's disparate impact. Calif. Law Rev. **104**, 671–732 (2016)
4. Bengtsson, L., Lu, X., Thorson, A., Garfield, R., Von Schreeb, J.: Improved response to disasters and outbreaks by tracking population movements with mobile phone network data: a post-earthquake geospatial study in Haiti. PLoS Med. **8**(8), e1001083 (2011)
5. Benkler, Y.: The Wealth of Networks. Yale University Press, New Haven (2006)
6. Berendt, B., Preibusch, S.: Better decision support through exploratory discrimination-aware data mining: foundations and empirical evidence. Artif. Intell. Law **22**(2), 1572–8382 (2014)
7. Blondel, V.D., Decuyper, A., Krings, G.: A survey of results on mobile phone datasets analysis. EPJ Data Sci. **4**(10) (2015)
8. Blumenstock, J., Cadamuro, G., On, R.: Predicting poverty and wealth from mobile phone metadata. Science **350**(6264), 1073–1076 (2015)
9. Bogomolov, A., Lepri, B., Ferron, M., Pianesi, F., Pentland, A.: Daily stress recognition from mobile phone data, weather conditions and individual traits. In: Proceedings of the 22nd ACM International Conference on Multimedia, pp. 477–486 (2014)
10. Bogomolov, A., Lepri, B., Staiano, J., Oliver, N., Pianesi, F., Pentland, A.: Once upon a crime: towards crime prediction from demographics and mobile data. In: Proceedings of the International Conference on Multimodal Interaction (ICMI), pp. 427–434 (2014)

11. Bogomolov, A., Lepri, B., Staiano, J., Letouzé, E., Oliver, N., Pianesi, F., Pentland, A.: Moves on the street: classifying crime hotspots using aggregated anonymized data on people dynamics. Big Data **3**(3), 148–158 (2015)
12. Burrell, J.: How the machine 'thinks': understanding opacity in machine learning algorithms. Big Data Soc. **3**(1) (2016)
13. Calders, T., Verwer, S.: Three naive Bayes approaches for discrimination-free classification. Data Min. Knowl. Disc. **21**(2), 277–292 (2010)
14. Calders, T., Zliobaite, I.: Why unbiased computational processes can lead to discriminative decision procedures. In: Custers, B., Calders, T., Schermer, B., Zarsky, T. (eds.) Discrimination and Privacy in the Information Society, pp. 43–57. Springer, Berlin (2013)
15. Centellegher, S., De Nadai, M., Caraviello, M., Leonardi, C., Vescovi, M., Ramadian, Y., Oliver, N., Pianesi, F., Pentland, A., Antonelli, F., Lepri, B.: The mobile territorial lab: a multilayered and dynamic view on parents' daily lives. EPJ Data Sci. **5**(3) (2016)
16. Chainey, S.P., Tompson, L., Uhlig, S.: The utility of hotspot mapping for predicting spatial patterns of crime. Secur. J. **21**, 4–28 (2008)
17. Christin, A., Rosenblatt, A., boyd, d.: Courts and predictive algorithms. Data Civil Rights Primer (2015)
18. Citron, D.K., Pasquale, F.: The scored society. Wash. Law Rev. **89**(1), 1–33 (2014)
19. Crawford, K., Schultz, J.: Big data and due process: toward a framework to redress predictive privacy harms. Boston College Law Rev. **55**(1), 93–128 (2014).
20. De Choudhury, M., Gamon, M., Counts, S., Horvitz, E.: Predicting depression via social media. In: Proceedings of the 7th International AAAI Conference on Weblogs and Social Media (2013)
21. De Filippi, P.: The interplay between decentralization and privacy: the case of blockchain technologies. J. Peer Production **7** (2015)
22. de Montjoye, Y.-A., Hidalgo, C., Verleysen, M., Blondel, V.: Unique in the crowd: the privacy bounds of human mobility. Sci. Rep. **3** (2013)
23. de Montjoye, Y.-A., Shmueli, E., Wang, S., Pentland, A.: OpenPDS: protecting the privacy of metadata through safeanswers. PLoS One **9**(7), e98790 (2014)
24. de Montjoye, Y.-A., Radaelli, L., Singh, V.K., Pentland, A.: Unique in the shopping mall: on the re-identifiability of credit card metadata. Science **347**(6221), 536–539 (2015)
25. de Oliveira, R., Karatzoglou, A., Concejero Cerezo, P., Armenta Lopez de Vicuña, A., Oliver, N.: Towards a psychographic user model from mobile phone usage. In: CHI'11 Extended Abstracts on Human Factors in Computing Systems, pp. 2191–2196. ACM, New York (2011)
26. Devarajan, S.: Africa's statistical tragedy. Rev. Income Wealth **59**(S1), S9–S15 (2013)
27. Diakopoulos, N.: Algorithmic accountability: journalistic investigation of computational power structures. Digit. Journal. **3**(3), 398–415 (2015)
28. Easterly, W.: The Tyranny of Experts. Basic Books, New York (2014)
29. Eck, J., Chainey, S., Cameron, J., Wilson, R.: Mapping crime: understanding hotspots. National Institute of Justice, Washington (2005)
30. Faurholt-Jepsena, M., Frostb, M., Vinberga, M., Christensena, E.M., Bardram, J.E., Kessinga, L.V.: Smartphone data as objective measures of bipolar disorder symptoms. Psychiatry Res. **217**, 124–127 (2014)
31. Feldman, M., Friedler, S.A., Moeller, J., Scheidegger, C., Venkatasubramanian, S.: Certifying and removing disparate impact. In: Proceedings of the 21th ACM SIGKDD International Conference on Knowledge Discovery and Data Mining, pp. 259–268 (2015)
32. Ferguson, A.G.: Crime mapping and the fourth amendment: redrawing high-crime areas. Hastings Law J. **63**, 179–232 (2012)
33. Fields, G.: Changes in poverty and inequality. World Bank Res. Obs. **4**, 167–186 (1989)
34. Fiske, S.T.: Stereotyping, prejudice, and discrimination. In: Gilbert, D.T., Fiske, S.T., Lindzey, G. (eds.) Handbook of Social Psychology, pp. 357–411. McGraw-Hill, Boston (1998)
35. Frias-Martinez, E., Williamson, G., Frias-Martinez, V.: An agent-based model of epidemic spread using human mobility and social network information. In: 2011 International Conference on Social Computing (SocialCom), pp. 57–64. IEEE, New York (2011)

36. Gillespie, T.: The relevance of algorithms. In: Gillespie, T., Boczkowski, P., Foot, K. (eds.) Media Technologies: Essays on Communication, Materiality, and Society, pp. 167–193. MIT Press, Cambridge (2014)
37. Ginsberg, J., Mohebbi, M.H., Patel, R.S., Brammer, L., Smolinski, M.S., Brilliant, L.: Detecting influenza epidemics using search engine query data. Nature **457**, 1012–1014 (2009)
38. Hajian, S., Bonchi, F., Castillo, C.: Algorithmic bias: from discrimination discovery to fairness-aware data mining. In: Proceedings of the 22nd ACM SIGKDD International Conference on Knowledge Discovery and Data Mining, pp. 2125–2126. ACM, New York (2016)
39. Jean, N., Burke, M., Xie, M., Davis, W.M., Lobell, D.B., Ermon, S.: Combining satellite imagery and machine learning to predict poverty. Science **353**(6301), 790–794 (2016)
40. Jerven, M.: Poor Numbers: How We Are Misled by African Development Statistics and What to Do About It. Cornell University Press, Ithaca (2013)
41. King, G.: Ensuring the data-rich future of the social sciences. Science **331**(6018), 719–721 (2011)
42. Kosinski, M., Stillwell, D., Graepel, T.: Private traits and attributes are predictable from digital records of human behavior. Proc. Natl. Acad. Sci. **110**(15), 5802–5805 (2013)
43. Kuznets, S.: Economic growth and income inequality. Am. Econ. Rev. **45**, 1–28 (1955)
44. Latzer, M., Hollnbuchner, K., Just, N., Saurwein, F.: The economics of algorithmic selection on the internet. In: Bauer, J., Latzer, M. (eds.) Handbook on the Economics of the Internet. Edward Elgar, Cheltenham (2015)
45. Lazer, D., Pentland, A., Adamic, L., Aral, S., Barabasi, A-L., Brewer, D., Christakis, N., Contractor, N., Fowler, J., Gutmann, M., Jebara, T., King, G., Macy, M., Roy, D., Van Alstyne, M.: Computational social science. Science **323**(5915), 721–723 (2009)
46. Lepri, B., Staiano, J., Shmueli, E., Pianesi, F., Pentland, A.: The role of personality in shaping social networks and mediating behavioral change. User Model. User-Adap. Inter. **26**(2), 143–175 (2016)
47. LiKamWa, R., Liu, Y., Lane, N.D., Zhong, L.: Moodscope: building a mood sensor from smartphone usage patterns. In: Proceedings of the 11th Annual International Conference on Mobile Systems, Applications, and Service (MobiSys), pp. 389–402 (2013)
48. Liu, H.Y., Skjetne, E., Kobernus, M.: Mobile phone tracking: in support of modelling traffic-related air pollution contribution to individual exposure and its implications for public health impact assessment. Environ. Health **12**, 93 (2013)
49. Louail, T., Lenormand, M., Cantu Ros, O.G., Picornell, M., Herranz, R., Frias-Martinez, E., Ramasco, J.J., Barthelemy, M.: From mobile phone data to the spatial structure of cities. Sci. Rep. **4**, 5276 (2014)
50. Lu, X., Bengtsson, L., Holme, P.: Predictability of population displacement after the 2010 haiti earthquake. Proc. Natl. Acad. Sci. **109**, 11576–11581 (2012)
51. Major, B., O'Brien, L.T.: The social psychology of stigma. Annu. Rev. Psychol. **56**, 393–421 (2005)
52. Matic, A., Oliver, N.: The untapped opportunity of mobile network data for mental health. In: Future of Pervasive Health Workshop, vol. 6. ACM, New York (2016)
53. Mohler, G.O., Short, M.B., Brantingham, P.J., Schoenberg, F.P., Tita, G.E.: Self-exciting point process modeling of crime. J. Am. Stat. Assoc. **106**, 100–108 (2011)
54. Nakamoto, S.: Bitcoin: a peer-to-peer electronic cash system. Technical Report, Kent University (2009)
55. Ofli, F., Meier, P., Imran, M., Castillo, C., Tuia, D., Rey, N., Briant, J., Millet, P., Reinhard, F., Parkan, M., Joost, S.: Combining human computing and machine learning to make sense of big (aerial) data for disaster response. Big Data **4**, 47–59 (2016)
56. Ohm, P.: Broken promises of privacy: responding to the surprising failure of anonymization. UCLA Law Rev. **57**, 1701–1777 (2010)
57. Oliver, N., Matic, A., Frias-Martinez, E.: Mobile network data for public health: opportunities and challenges. Front. Public Health **3**, 189 (2015)

58. O'Neil, C.: Weapons of Math Destruction: How Big Data Increases Inequality and Threatens Democracy. Crown, New York (2016).
59. Osmani, V., Gruenerbl, A., Bahle, G., Lukowicz, P., Haring, C., Mayora, O.: Smartphones in mental health: detecting depressive and manic episodes. IEEE Pervasive Comput. **14**(3), 10–13 (2015)
60. Pager, D., Shepherd, H.: The sociology of discrimination: racial discrimination in employment, housing, credit and consumer market. Annu. Rev. Sociol. **34**, 181–209 (2008)
61. Pasquale, F.: The Black Blox Society: The Secret algorithms That Control Money and Information. Harvard University Press, Cambridge (2015)
62. Pastor-Escuredo, D., Torres Fernandez, Y., Bauer, J.M., Wadhwa, A., Castro-Correa, C., Romanoff, L., Lee, J.G., Rutherford, A., Frias-Martinez, V., Oliver, N., Frias-Martinez, E., Luengo-Oroz, M.: Flooding through the lens of mobile phone activity. In: IEEE Global Humanitarian Technology Conference, GHTC'14. IEEE, New York (2014)
63. Pentland, A.: Society's nervous system: building effective government, energy, and public health systems. IEEE Comput. **45**(1), 31–38 (2012)
64. Perry, W.L., McInnis, B., Price, C.C., Smith, S.C., Hollywood, J.S.: Predictive Policing: The Role of Crime Forecasting in Law Enforcement Operations. Rand Corporation, Santa Monica (2013)
65. Podesta, J., Pritzker, P., Moniz, E.J., Holdren, J., Zients, J.: Big data: seizing opportunities, preserving values. Technical Report, Executive Office of the President (2014)
66. Ramirez, E., Brill, J., Ohlhausen, M.K., McSweeny, T.: Big data: a tool for inclusion or exclusion? Technical Report, Federal Trade Commission, January 2016
67. Ratcliffe, J.H.: A temporal constraint theory to explain opportunity-based spatial offending patterns. J. Res. Crime Delinq. **43**(3), 261–291 (2006)
68. Ravallion, M.: The economics of poverty: history, measurement, and policy. Oxford University Press, Oxford (2016)
69. Ribeiro, M.T., Singh, S., Guestrin, C.: "why should I trust you?": explaining the predictions of any classifier. In: Proceedings of the 22nd ACM SIGKDD International Conference on Knowledge Discovery and Data Mining, San Francisco, CA, USA, August 13–17, 2016, pp. 1135–1144 (2016)
70. Samuelson, W., Zeckhauser, R.: Status quo bias in decision making. J. Risk Uncertain. **1**, 7–59 (1988)
71. San Pedro, J., Proserpio, D., Oliver, N.: Mobiscore: towards universal credit scoring from mobile phone data. In: Proceedings of the International Conference on User Modeling, Adaptation and Personalization (UMAP), pp. 195–207 (2015)
72. Short, M.B., D'Orsogna, M.R., Pasour, V.B., Tita, G.E., Brantingham, P.J., Bertozzi, A.L., Chayes, L.B.: A statistical model of criminal behavior. Math. Models Methods Appl. Sci. **18**(supp01), 1249–1267 (2008)
73. Singh, V.K., Freeman, L., Lepri, B., Pentland, A.: Predicting spending behavior using socio-mobile features. In: 2013 International Conference on Social Computing (SocialCom), pp. 174–179. IEEE, New York (2013)
74. Singh, V.K., Bozkaya, B., Pentland, A.: Money walks: implicit mobility behavior and financial well-being. PLoS One **10**(8), e0136628 (2015)
75. Smith-Clarke, C., Mashhadi, A., Capra, L.: Poverty on the cheap: estimating poverty maps using aggregated mobile communication networks. In: Proceedings of the 32nd ACM Conference on Human Factors in Computing Systems (CHI2014) (2014)
76. Soto, V., Frias-Martinez, V., Virseda, J., Frias-Martinez, E.: Prediction of socioeconomic levels using cell phone records. In: Proceedings of the International Conference on UMAP, pp. 377–388 (2011)
77. Staiano, J., Oliver, N., Lepri, B., de Oliveira, R., Caraviello, M., Sebe, N.: Money walks: a human-centric study on the economics of personal mobile data. In: Proceedings of the 2014 ACM International Joint Conference on Pervasive and Ubiquitous Computing, pp. 583–594. ACM, New York (2014)

78. Staiano, J., Zyskind, G., Lepri, B., Oliver, N., Pentland, A.: The rise of decentralized personal data markets. In: Shrier, D., Pentland, A. (eds.) Trust::Data: A New Framework for Identity and Data Sharing. CreateSpace Independent Publishing Platform (2016)
79. Sweeney, L.: Discrimination in online ad delivery. Available at SSRN: http://ssrn.com/abstract=2208240 (2013)
80. Szabo, N.: Formalizing and securing relationships on public networks. First Monday **2**(9) (1997)
81. Thomas, L.: Consumer Credit Models: Pricing, Profit, and Portfolios. Oxford University Press, New York (2009)
82. Tizzoni, M., Bajardi, P., Decuyper, A., Kon Kam King, G., Schneider, C.M., Blondel, V., Smoreda, Z., Gonzalez, M.C., Colizza, V.: On the use of human mobility proxies for modeling epidemics. PLoS Comput. Biol. **10**(7) (2014)
83. Tobler, C.: Limits and potential of the concept of indirect discrimination. Technical Report, European Network of Legal Experts in Anti-Discrimination (2008)
84. Toole, J.L., Eagle, N., Plotkin, J.B.: Spatiotemporal correlations in criminal offense records. ACM Trans. Intell. Syst. Technol. **2**(4), 38:1–38:18 (2011)
85. Traunmueller, M., Quattrone, G., Capra, L.: Mining mobile phone data to investigate urban crime theories at scale. In: Proceedings of the International Conference on Social Informatics, pp. 396–411 (2014)
86. Tufekci, Z.: Algorithmic harms beyond Facebook and Google: emergent challenges of computational agency. Colorado Technol. Law J. **13**, 203–218 (2015)
87. Tverksy, A., Kahnemann, D.: Judgment under uncertainty: heuristics and biases. Science **185**(4157), 1124–1131 (1974)
88. Venerandi, A., Quattrone, G., Capra, L., Quercia, D., Saez-Trumper, D.: Measuring urban deprivation from user generated content. In: Proceedings of the 18th ACM Conference on Computer Supported Cooperative Work & Social Computing (CSCW2015) (2015)
89. Vescovi, M., Perentis, C., Leonardi, C., Lepri, B., Moiso, C.: My data store: Toward user awareness and control on personal data. In: Proceedings of the 2014 ACM International Joint Conference on Pervasive and Ubiquitous Computing: Adjunct Publication, pp. 179–182 (2014)
90. Wang, T., Rudin, C., Wagner, D., Sevieri, R.: Learning to detect patterns of crime. In: Machine Learning and Knowledge Discovery in Databases, pp. 515–530. Springer, Berlin (2013)
91. Wang, H., Li, Z., Kifer, D., Graif, C.: Crime rate inference with big data. In: Proceedings of International Conference on KDD (2016)
92. Want, R., Pering, T., Danneels, G., Kumar, M., Sundar, M., Light, J.: The personal server: changing the way we think about ubiquitous computing. In: Proceedings of 4th International Conference on Ubiquitous Computing, pp. 194–209 (2002)
93. Weisburd, D.: Place-based policing. Ideas Am. Policing **9**, 1–16 (2008)
94. Wesolowski, A., Eagle, N., Tatem, A., Smith, D., Noor, R., Buckee, C.: Quantifying the impact of human mobility on malaria. Science **338**(6104), 267–270 (2012)
95. Wesolowski, A., Stresman, G., Eagle, N., Stevenson, J., Owaga, C., Marube, E., Bousema, T., Drakeley, C., Cox, J., Buckee, C.O.: Quantifying travel behavior for infectious disease research: a comparison of data from surveys and mobile phones. Sci. Rep. **4** (2014)
96. Willson, M.: Algorithms (and the) everyday. Inf. Commun. Soc. **20**, 137–150 (2017)
97. Wilson, R., Erbach-Schoenengerg, E., Albert, M., Power, D., Tudge, S., Gonzalez, M., et al.: Rapid and near real-time assessments of population displacement using mobile phone data following disasters: the 2015 Nepal earthquake. PLoS Current Disasters, February 2016
98. Zang, H., Bolot, J.: Anonymization of location data does not work: a large-scale measurement study. In: Proceedings of 17th ACM Annual International Conference on Mobile Computing and Networking, pp. 145–156 (2011)
99. Zarsky, T.Z.: Automated prediction: Perception, law and policy. Commun. ACM **4**, 167–186 (1989)

100. Zarsky, T.: The trouble with algorithmic decisions: an analytic road map to examine efficiency and fairness in automated and opaque decision making. Sci. Technol. Hum. Values **41**(1), 118–132 (2016)
101. Zyskind, G., Nathan, O., Pentland, A.: Decentralizing privacy: using blockchain to protect personal data. In: Proceedings of IEEE Symposium on Security and Privacy Workshops, pp. 180–184 (2014)

Enabling Accountability of Algorithmic Media: Transparency as a Constructive and Critical Lens

Nicholas Diakopoulos

Abstract As the news media adopts opaque algorithmic components into the production of news information it raises the question of how to maintain an accountable media system. One practical mechanism that can help expose the journalistic process, algorithmic or otherwise, is transparency. Algorithmic transparency can help to enable media accountability but is in its infancy and must be studied to understand how it can be employed in a productive and meaningful way in light of concerns over user experience, costs, manipulation, and privacy or legal issues. This chapter explores the application of an algorithmic transparency model that enumerates a range of possible information to disclose about algorithms in use in the news media. It applies this model as both a constructive tool, for guiding transparency around a news bot, and as a critical tool for questioning and evaluating the disclosures around a computational news product and a journalistic investigation involving statistical inferences. These case studies demonstrate the utility of the transparency model but also expose areas for future research.

Abbreviations

CAR	Computer-assisted reporting
NPR	National Public Radio
RTDNA	Radio Television Digital News Association
SPJ	Society for Professional Journalists
SRF	Schweizer Radio und Fernsehen

N. Diakopoulos (✉)
Philip Merrill College of Journalism, University of Maryland, College Park, Knight Hall, 7765 Alumni Drive, College Park, MD 20742, USA
e-mail: nad@umd.edu

T. Cerquitelli et al. (eds.), *Transparent Data Mining for Big and Small Data*,
Studies in Big Data 32, DOI 10.1007/978-3-319-54024-5_2

1 Introduction

The news media is rapidly adopting algorithmic approaches to the underlying journalistic tasks of information collection, curation, verification, and dissemination. These algorithmic approaches are changing the production pipeline of news as well as the user experience of news consumption. Major news outlets like the Associated Press routinely use automated writing software to produce corporate earnings articles based on structured data [1], or deploy news bots that disseminate and expand the audience for news information [2]. Statistical models can now be found simulating, predicting, or visualizing news stories in domains like public health or elections.[1] Recommendation systems steer users towards popular news articles based on social activity on Twitter or Facebook, or on content consumption patterns [3]. Headlines written by editors are optimized through A/B tests, or even algorithmically generated from scratch [4]. Breaking news alerts are accelerated through story clustering and detection [5], or by developing statistical techniques that can alert journalists to potential emerging stories [6]. And social listening and curation tools are used to identify, verify, and rank social media content as it finds its way to the homepage [7, 8].

Questions about media accountability and ethics emerge as these algorithmic approaches become more entrenched in how the media system produces and mediates news information. While such technologies enable scale and an ostensibly objective and factual approach to editorial decision making, they also harbor biases in the information that they include, exclude, highlight, or make salient to journalists or end-users. Such biases can be consequential for the formation of publics and the fair and uniform provision of information [9]. At the same time, such black box systems can be difficult to comprehend due to the opacity in their automated decision-making capabilities [10, 11].

Recently media ethics in journalism has seen a shift away from objectivity as a sole focus. The key issue with objectivity, as media ethicist Stephen Ward puts it, is "If a news report involves (at least some) interpretation, how can it be objective?" [12]. Pervasive interpretation in news making problematizes objectivity, from deciding what is newsworthy, to choosing the words and frames of a story, to amplifying the salience of some elements over others in a visual design. The norm of objectivity is waning in the face of substantial growth of what some have termed "contextual journalism" [13]—that is news that advances analysis and interpretation via context. In its stead there is an increasing emphasis on transparency as the guiding norm [14], which is now listed prominently in several ethics codes such as

[1] http://www.nytimes.com/interactive/2016/upshot/presidential-polls-forecast.html.

that of the Society for Professional Journalists (SPJ),[2] the Radio Television Digital News Association (RTDNA),[3] and National Public Radio (NPR).[4]

There is certainly an aspect of algorithms that are objective: they execute the same set of procedures consistently each time they are run. And the corporate rhetoric surrounding algorithms often frames them as objective and devoid of human influence. But researchers have long criticized the embedded biases in algorithms, which can emerge through their human design or the data that feeds and trains them [15]. Much as human decision processes are subject to individual or group biases, algorithms also harbor bias. In some cases algorithms are much more explicit about bias (e.g. if embedded rules or definitions are at play), and other times algorithmic bias is deeply mediated via machine learning processes that learn biases from the training data they are fed. There is a parallel here: algorithms also "interpret" the information they process via the biases designed into or emergent in them, and thus the same skepticism of objectivity brought against traditional journalism is also applicable to algorithmic media. Transparency as a normative practice to facilitate media accountability can similarly be employed as way to verify the interpretation of an algorithmic component used in the newsmaking process.

It's worth noting that transparency is not a silver bullet for media ethics; a range of other media accountability approaches like press councils, clear codes of conduct, and healthy media criticism are all useful and appropriate in different contexts [16]. Transparency is valuable for media accountability because it facilitates testing and validating the interpretations that underlie newswork, traditional or algorithmic. Deuze [17] defines transparency as the "ways in which people both inside and external to journalism are given a chance to monitor, check, criticize and even intervene in the journalistic process." Transparency entails the disclosure of information about the news production process so that audiences can monitor that process, which reinforces the legitimacy of the media.

Transparency around newswork has traditionally involved strategies like linking to sources and documents used in reporting, signaling source expertise and qualifications to comment on a given topic, offering a window into the editorial decision making process, noting relationships with partners or funders, or disclosing and correcting errors or failures [18]. But with algorithms and data it's less clear how to translate the ideal of transparency espoused in ethics guides into practical strategies for media organizations. The goal of this chapter is to begin to fill this gap in pragmatic understanding of how to engage in effective algorithmic transparency.

Drawing on prior research, this chapter will first outline a set of information that might be disclosed about the types of algorithms in use at news media organizations. It is important to understand the space of possible information that might be disclosed about an algorithm as part of a transparency effort. Then this model of algorithmic transparency will be illustrated using case studies in

---

[2] http://www.spj.org/ethicscode.asp.

[3] http://www.rtdna.org/content/rtdna_code_of_ethics.

[4] http://ethics.npr.org/.

two capacities: constructive and critical. First, the model will be used as a way to structure and think constructively about algorithmic transparency in a news automation bot that we built called AnecbotalNYT. Then the model will be used as a critical lens to enable an evaluation of both editorial as well as news information products from Google and BuzzFeed News. Finally, the chapter will conclude with a discussion of challenges in applying algorithmic transparency in practice as well as opportunities for further research.

2 Transparency Model

This section presents a model of algorithmic transparency consisting of an enumeration of information factors that might be disclosed about algorithms in use in the news media. The full description of the model and how it was developed is presented in [19] but is briefly summarized here so that it's clear how the model is being applied to the case studies presented later in this chapter.

The model includes a set of pragmatic dimensions of information that might be disclosed as part of an algorithmic transparency effort. The dimensions of information disclosure were elicited through a series of focus groups that asked groups of 10–15 scholars and practitioners knowledgeable in algorithmic news media to brainstorm and discuss what might be disclosed around three case studies of algorithms used in news production. The cases were selected to address diverse aspects of how algorithms are used in news production processes and included automated news writing (e.g. data-driven writing software), algorithmic curation (e.g. moderation or recommendation), and simulation, prediction, and modeling in storytelling (e.g. election forecasts, visualized interactive models). Moderator-led focus groups spent an hour considering each of the cases and generated a host of candidate information factors that might be made transparent about the algorithms at the focus of that case. The transcripts for these focus groups were qualitatively analyzed through a process of iterative coding, affinity diagramming, typologizing and memoing [20] and indicated a set of layers across which information about an algorithm could be made transparent.

The four layers of information disclosure that were identified include: data, model, inference, and interface. Individual aspects that were discussed in the focus groups across the four layers articulated are shown in Table 1. The *data* layer is a key aspect where transparency is needed in algorithmic systems, particularly those that rely on machine learning. The information quality and validity of the data feeding algorithms was seen as paramount: garbage in, garbage out. Quality includes aspects of accuracy, uncertainty, timeliness, comprehensiveness, and provenance, as well as how data has been transformed, validated, cleaned, or edited. The *model* process and methodology for modeling were other aspects deemed worthy of disclosure. Details of the model might include the features, variables, weights, and type of model used. A model might also include heuristics, assumptions, rules, or constraints that might be useful and helpful to disclose. The output *inferences* from

Table 1 The transparency model used throughout this paper includes aspects of disclosable information across four layers: data, model, inference, interface

Layer	Factors
Data	• Information quality o Accuracy o Uncertainty (e.g. error margins) o Timeliness o Completeness • Sampling method • Definitions of variables • Provenance (e.g. sources, public or private) • Volume of training data used in machine learning • Assumptions of data collection • Inclusion of personally identifiable information
Model	• Input variables and features • Target variable(s) for optimization • Feature weightings • Name or type of model • Software modeling tools used • Source code or pseudo-code • Ongoing human influence and updates • Explicitly embedded rules (e.g. thresholds)
Inference	• Existence and types of inferences made • Benchmarks for accuracy • Error analysis (including e.g. remediation standards) • Confidence values or other uncertainty information
Interface	• Algorithm "in use" notification • On/off • Tweakability of inputs, weights

an algorithmic process such as the classifications, predictions, or recommendations produced raised issues around errors, uncertainty, and accuracy. Transparency information might include what types of inferences are being made, as well as error analysis (and remediation standards for dealing with remaining errors), and disclosure of confidence values on inferences. Finally, at the interface layer we can talk about aspects of how algorithmic transparency information is integrated into the user experience, including through deeper interactivity, perturbation or sensitivity analysis, or even capabilities to override the algorithms (e.g. an on/off switch).

The results of the study also emphasize the deep entanglements of humans within algorithmic systems and the many human decisions that may need to be articulated as essential context in an algorithmic transparency effort. Such transparency around human involvement might include disclosing aspects of editorial goals or rationale for selection, inclusion, or exclusion of various inputs and outputs to an algorithm, disclosing details of the configuration or parameterization of the system, and talking about adjustments to the system over time. In addition, knowing about the responsible actors, including the authors or designers of such systems could enhance the personal accountability and responsibility that individuals feel for such systems.

Barriers and challenges to effective algorithm transparency efforts were discussed. These included: concerns over disclosure of *proprietary* information that would damage competitive advantages, or leave a system open to *manipulation*; concerns with *overwhelming users* with too much information in the user interface that they might tune-out or not find relevant; business issues surrounding the *costs* of data preparation, documentation, or benchmark testing that would be entailed by a transparency effort; *privacy* considerations from disclosure of improperly anonymized data; and *legal* implications of admitting error or uncertainty in decision processes. Several of these issues, such as proprietary concerns and complexity in information presentation, are echoed in other research on the opacity of algorithms [21].

Subsequent sections in this chapter will present applications of this transparency model to specific instances of the use of algorithms in the news media. Different contexts and use cases may require the application of different subsets of facets enumerated in Table 1. The case studies take a user-centric approach in trying to align information with the decisions of end-users so that disclosures about algorithms maximize the possibility that users respond to that information [22]. Any information disclosure should have the potential to either impact an individual user's decision processes, or wider public understanding of aggregate system behavior. For that reason, during the application of the model to case studies, questions relating to the user decisions and the impacts of information disclosure that affect those decisions will be considered:

- What are the decisions the user or wider public would make based on the algorithm in question?
- What are the relevant bits of information about the algorithm that would help users or the public make those decisions more effectively? How might users respond to this information?
- What's the worst case outcome if users can't make such decisions effectively?

Moreover, the barriers and challenges to transparency that the model articulates will be considered. For instance, the issues of gaming, manipulation, and circumvention as well as the consequences of those actions will be examined. Aspects of how different types of information disclosures affect the user experience, as well as the costs of transparency information production will be taken into account.

3 Case Studies

In this section the model introduced above will be applied in several case studies. First, a computational journalism project involving the creation of a news bot will be described, including the ways in which the model informed the editorial transparency of the algorithm used by the bot. Then, two case studies will be presented in which the model is used to develop critiques of a computational news product and a journalistic investigation.

3.1 Editorial Transparency in an Automated Newsbot

News bots are automated agents that participate in news and information dissemination, often on social media platforms like Twitter or in chat applications like Facebook Messenger [2]. Such bots can serve a range of journalistic tasks from curating and aggregating sources of information, to analyzing and processing data, and bridging different platforms to raise awareness or further dissemination. As newsrooms seek to develop and build more of these bots and other information appliances, it raises the question of how to express and provide editorial transparency of such tools.

Here we consider a Twitter news bot called AnecbotalNYT that we built to help surface and raise awareness for interesting personal experiences or anecdotes that relate to news articles published by the New York Times [23]. AnecbotalNYT works by listening to the public Twitter stream for anyone sharing a New York Times article link. It then collects all of the comments that are published on that article via the New York Times Community API, scores those comments according to several factors including length, reading level, and personal experience [8], and then ranks and selects a comment that is likely to contain a personal story or anecdote. Finally, the bot responds to the person who originally shared the article link with an image that contains the text of the comment. An example of the bot responding is shown in Fig. 1.

Fig. 1 An example of the output produced by the AnecbotalNYT Twitter bot

There are numerous editorial decisions that are embedded in the functioning of the bot, from where the data comes from, and how it's sampled, to how comments are filtered, scored, and ranked, as well as thresholds used to define reasonable constraints that improve the end-user experience of the bot such as filtering out short comments and ceasing to tweet if API limits are reached. In order to provide transparency for these various factors we were informed by the algorithmic transparency model presented above as we developed and published transparency information on the bot's Github page.[5] As we considered what transparency information to include on the page we worked through the potential factors for disclosure enumerated in Table 1, asking whether there might be interest from potential end-users in having that information.

As a result, the Github page includes not only the full code base for the bot, as well as descriptions of the data, model, and inferences that the bot makes, but it also links to context such as research papers that informed the bot's design. Moreover, we explicitly parameterized many of the editorial rules the bot follows in an effort to make them explicit and salient. Instead of having to parse through dense code to see these editorial rules, they are embedded within configuration files which we felt would be easier to read and understand for non-technical users. Of course not everything could or should be parameterized but we identified what we thought were the most subjective editorial rules like the weights applied to the individual subscores in aggregating a final score, the threshold used to filter out comments that were short, or even how often the bot would potentially respond.

This approach to algorithmic transparency of the bot was motivated with two possible sets of users in mind: professional journalists who might adapt the open source code, and Twitter users who encounter the bot or had the bot respond to their tweets.

For professionals, the transparency was motivated by the desire to make sure that news organizations that might want to build on the open source code could explicitly see the design decisions embedded in the bot's algorithm and thus more clearly edit and produce the desired output according to their own editorial goals. The decision that such a news organization needs to make is whether it will adapt and use the open source code of the bot. The transparency information disclosed in the Github profile thus facilitates the decision making process by showing the code as well as showing how the bot makes decisions and can be flexibly re-parameterized for different editorial uses.

On the other hand, for Twitter end-users, the bot provides transparency via its Twitter profile, which explicitly links to its Github page. Users' curiosity may be aroused after interacting with the bot and thus they may be interested to explore what the bot is, who made it, and why it behaves the way it does. For instance, why did the bot choose a particular comment to share, or why didn't it respond to a different link the user had shared? The decision the user is implicitly making when they interact with the bot is whether to trust or rely on the information being presented by the

[5] https://github.com/comp-journalism/Comment-Bot.

bot. Is the comment that the bot shared real and is it a legitimate commentary made with respect to the news article? The transparency information becomes a tool for end-users who interact with the bot to gain additional information and an account of the bot's behavior, thus facilitating a more satisfying user experience.

3.2 *Critical Application of the Algorithmic Transparency Model*

The algorithmic transparency model presented above can be used not only as a way to guide the editorial transparency of news-oriented automation, like bots, but also as a way to inform the critical analysis of algorithms throughout society, including those that impact the media system which is the focus in this chapter. This evokes the notion of algorithmic accountability [11] which often involves different methods, such as reverse engineering or auditing, but which also encompasses the approach of critical commentary taken here. In this sub-section I apply the model in two separate cases that explore criticism of an algorithmic system with little to no transparency, and of an algorithmic methodology with a high-level of transparency. The two cases respectively focus on a news information product that Google created to curate and present issue-oriented quotes from presidential candidates directly within search results, and a piece of computational investigative journalism produced by Buzzfeed News in which simulation was used to identify potential fraud in professional Tennis.

3.2.1 Google's Issue Guide

On February 1st, 2016 Google launched a feature on its search results that highlighted US presidential candidates' views on various political issues like abortion, immigration, or taxes. If a user searched for "Hillary Clinton abortion" their search results would include an infobox generated by Google at the top of the page in a prominent position. If the candidate had provided a statement on the issue it was presented first, and then the user could browse through quotes by the presidential candidate extracted from news articles around the political issue. A view of search results showing all of the issues and with a specific issue (Economy and Jobs) expanded is shown in Fig. 2. Our original analysis and assessment of the bias in the tool was published in journalistic form by Slate [24].

Google does not position itself in the market as a media company but rather as a technology company. A common argument that platforms like Google, Twitter, and Facebook use is that since they do not produce original content, they should not be considered media companies [25]. However, in this case, Google is taking on a similar role to that of a news organization in terms making editorial decisions about the selection and salience of information that is available when users search for

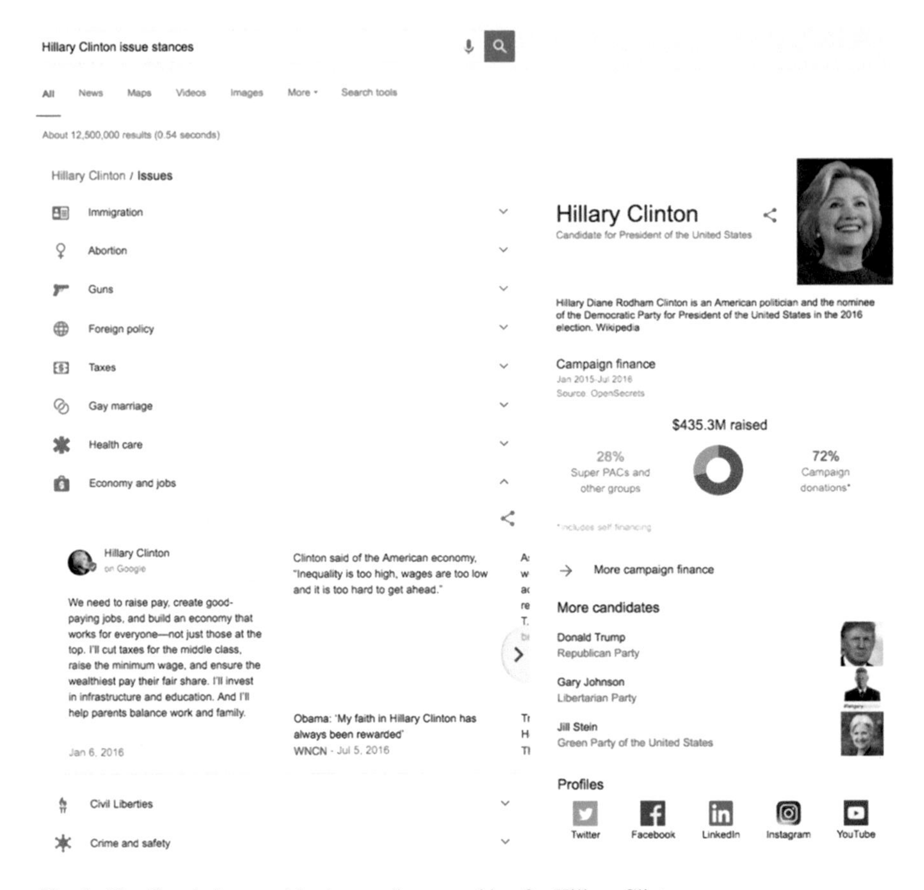

Fig. 2 The Google issue guide shown when searching for Hillary Clinton

candidate information. This matters because search engines are powerful arbiters of human attention that have the potential to shape the electorate's attention and voting propensities [26] and are seen as trustworthy by a majority (73% according to a 2012 Pew study) of users [27].

To examine the biases that are visible in this new Google feature we gathered data including all of the news articles and quotes that were selected for the infoboxes on April 1, 2016 and again on April 22nd and May 13th, 2016. Our approach towards analyzing the data collected and asking questions based on our observations was informed by the algorithmic transparency model above.

The first aspect of the tool that we observed was that the ordering of the issues presented by the guide is not alphabetical, which might be the simplest logical algorithm for sorting the list of issues. The ordering bias of the issues is consequential because items listed higher up on search results pages get more attention from searchers [28]. We then considered whether that ranking might come from issue importance polls conducted by Gallup and Pew but found no correlation.

Because it wasn't clear what input variables and features were being used to rank the issues, this informed a question that we posed to the Google representative that we spoke to about our investigation. According to that representative the issues are ranked using "several factors" which include user interest in the various issues as well as the ways that other organizations rank issues (e.g. Pew Vote Smart, and OnTheIssues.org were mentioned explicitly). Thus we had limited success in getting enough detail in the information that Google disclosed. Because the Google spokesperson mentioned user interest as a factor in the ranking we attempted to recreate the issue ordering using Google Trends data, which indicates the frequency of searches for specific terms over time. However we were unable to recreate the issue ordering using simple searches for the issue terms on Google Trends.

Another aspect of the investigation was whether there was evidence of human influence in the ranking of issues. We collected data three times over the course of about 6 weeks to understand if there were significant shifts, but found the list instead to be relatively static. This suggests (but of course does not confirm) that any data-driven techniques used to order the issues are being passed through a human editorial or review process. The only change that we observed in the 6 weeks of observation was that a new top-level issue was added: gay marriage. But in examining the search volume of "gay marriage" on Google trends it was clear that there was not a data-driven peak or spike in interest in the topic. Before the change, quotes referring to same-sex marriage were including under the "civil liberties" issue. Only in the data collected on May 13th were those quotes moved to a separate issue. This suggests that human editors decided at some point that this issue was important or distinct enough to warrant its own top-level category.

Another algorithmic (or semi-algorithmic) process presumed to be a part of the issue guide is how the quotes are selected and presented for each candidate. We confirmed the presence of an algorithmic process to select these quotes with the Google representative who said, "The quotes are algorithmically aggregated from news articles on Google Search." This transparency about the system thus checked boxes both about an algorithmic process "in use", as well as the provenance of the input data feeding the quote extraction (i.e. news articles returned via Google Search). The quote selection process operates by scanning news articles to identify quotes made by a candidate and then associates each quote found with one of the issues that are being tracked.

What was journalistically interesting about these illuminations was both the focus on quotes to begin with, which weights the importance of what candidates say versus what they do, as well as the biases that we observed in the rate at which quotes were found when comparing Clinton and Trump. It turns out that as of May 14th, 2016 Clinton had 299 quotes across all issues whereas Trump had only 160. Such a discrepancy is difficult to explain from an external perspective as bias could be entering the system in several ways. For instance, the discrepancy in the number of quotes for each candidate might be due to Google not indexing the same number of articles for each candidate, or because the candidates simply aren't being covered equally by the mainstream media. Informed by the transparency model we considered whether there might be an input sample bias that could be causing the output bias that we observed. To assess this we searched Google News for each candidate to enumerate all of the news articles that Google was indexing over the course of 1 month. We found 710 articles about Hillary Clinton, and 941 about Donald Trump, indicating that Trump is not underrepresented in the world of news that Google indexes. Other analyses have subsequently found a similar pattern in which Trump garners more media attention than Clinton.[6]

In this particular case we were interested in investigating the biases that were observable in the Google Issue Guide. The transparency model was informative for guiding certain questions that could be directed towards the company, such as confirming the use of algorithms, and eliciting information about the input variables and data. We felt that answers to these questions could inform a broad set of decisions that end-users might make based on the guide, such as whether to trust the information presented there, and whether the algorithm was treating the candidates fairly. By publishing this information we surmised that end-users might respond by discounting the tool since it does not clearly explain the biases that we observed. The algorithmic transparency model also guided the investigation by suggesting questions about input sampling bias that might be leading to the output biases observed, as well as orienting the investigation to look for signatures of human influence. While not yet at the stage of a "cookbook" that can guide the investigation or critique of an algorithm, we still found the model to be helpful in thinking through the *types* of information that we might be able to elicit or observe about the algorithm.

3.2.2 BuzzFeed News Investigation

Original analysis of data in a journalistic context is traditionally covered under the general rubric of Computer-Assisted Reporting (CAR) and has been practiced for decades [29]. It is not uncommon for news organizations that engage in such work to publish methodology articles or even whitepapers [30] that describe the process of analysis in a way that might be familiar to scientists who must be transparent about their data and methods for the sake of reproducibility. Outlets such as National Public Radio (NPR) and Schweizer Radio und Fernsehen (SRF) espouse the publication of open-source code and the use of programmatic data analysis and transformation in their data-journalism projects in order to ensure replicability.[7,8] More recently, increasingly sophisticated uses of computing that go beyond standard statistical analysis, such as through advanced modeling, simulation, or prediction have been used in journalistic contexts. News organizations such as the New York Times, FiveThirtyEight, BuzzFeed and many others[9] maintain repositories on

[6]http://thedataface.com/trump-media-analysis/.

[7]http://blog.apps.npr.org/2014/09/02/reusable-data-processing.html.

[8]https://github.com/grssnbchr/rddj-reproducibility-workflow.

[9]https://github.com/silva-shih/open-journalism.

Github, where they open-source their data and code in many (but not all) data- or computationally-driven news stories. In this section I will utilize the algorithmic transparency model as a mechanism to inform a critique of a particular case of a news organization being transparent in their use of simulation within a news investigation. The model is used to expose gaps in the methodological transparency offered in the project.

The focus of this case study is an article entitled "The Tennis Racket" published by BuzzFeed News [31]. This story investigates the broader issue of fraud in professional tennis, specifically as it relates to the possibility that players can easily throw a match (i.e. lose intentionally) and thus cause people betting against them to win. The sport is plagued by a lack of dedicated attention to both identifying and vigorously pursuing instances of suspected fraud, thus making it an interesting area for public interest investigation. The BuzzFeed investigation was particularly innovative in its use of simulation to identify professional players whose win-loss records in conjunction with betting odds statistics suggest that there was a suspiciously high chance they lost a game on purpose. By analyzing betting odds at both the beginning and ending of a match it's possible to identify instances where there was a large swing (e.g. greater than 10% difference) which might be considered unlikely (if we are to believe the bookmakers' initial odds calculation). If there are enough of these matches it casts suspicion on the player, but the article is clear to state that this statistical evidence is only suggestive and that "Identifying whether someone has fixed a match is incredibly difficult without betting records, without access to telephone records, without access to financial transactions." Thus, the names of suspicious players are not included in the article, or in the other transparency disclosures of the project.

An innovative aspect of the approach towards editorial transparency in this project is that transparency disclosures occur in the interface at three levels (T1, T2, T3) of increasing sophistication and detail and are targeted at different types of end-users. Linked from the original story is an article (T1) that can only be described as "Buzzfeed-y" in the sense that it includes comical animated.gifs as well as crystallized sub-heads and short accessible paragraphs that describe information about the methods and data used, as well as what was found as a result [32]. From there, a reader can click deeper on the "detailed methodology" (T2) which is displayed on a Github page[10] that describes six methodological details such as the data acquisition and preparation as well as other calculations. This detailed methodology then links to the final, and most detailed transparency disclosure (T3): the source code used to run the analysis which is presented as a Jupyter Notebook with python code and additional annotations describing what each section of code does in terms of the final analysis.[11] Each level of disclosure adds additional details and nuance to the understanding of the analysis and thus allows different

[10]https://github.com/BuzzFeedNews/2016-01-tennis-betting-analysis.

[11]https://github.com/BuzzFeedNews/2016-01-tennis-betting-analysis/blob/master/notebooks/tennis-analysis.ipynb.

stakeholders to access the granularity of information that is most relevant to their interests and needs. A truly dedicated journalist, or a professional tennis investigator might be motivated to use T3 information to rerun the analysis code for the sake of reproducibility, or to try it out with a different set of public or private data. But someone who is merely curious about the provenance and sample of the data could access the T2 description without the additional cognitive complexity of understanding code in T3. This approach to transparency disclosure follows a multi-level "pyramid" structure involving progressively denser and more detailed transparency information [19]. Importantly, it mitigates one of the primary concerns around implementing algorithmic transparency—that it is too difficult for end-users to make sense of transparency disclosures around algorithms.

The transparency disclosures across T1, T2, and T3 hit on many of the possible dimensions of disclosable information enumerated in the algorithmic transparency model across layers of data, model, and inference. But the model also suggests gaps where there could have been disclosures elaborated more precisely.

At the data layer, provenance is indicated (i.e. OddsPortal.com), as well as the purposive sampling strategy used to focus on certain types of matches (ATP and Grand Slam). At the same time, the sample was underspecified because the bookmakers from which the data was sampled were not named. All that is disclosed is that the data came from, "seven large, independent bookmakers whose odds are available on OddsPortal.com," but there were over 80 different bookmakers listed on OddsPortal.com when I looked in September 2016. If it was important to keep these bookmakers anonymous the disclosure could have included a rationale for that decision. This lack of sampling detail makes it more difficult to reproduce the analysis by re-collecting the data independently. Furthermore there is lack of specificity in the timeframe of data collected. We're told that the tennis matches analyzed occurred, "between 2009 and mid-September 2015," though the disclosure easily could have specified the exact dates so that their sample could be reproduced. Finally, there is at least one assumption around the data, which is that the bookmakers' initial odds are somehow accurate. Since the model relies on a shift in those odds the accuracy of the initial value is particularly important.

The model specification included in T2 and T3 is quite extensive. We are linked to the source code in T3 permitting very detailed examination of how the model was created and simulated. The target variable of odds-movement is clearly identified and defined. And we're explicitly told the specific thresholds used (i.e. 10%) to define what constituted a suspicious level in terms of odds-movement, which is further rationalized by citing "discussions with sports-betting investigators." However, the expertise and identity of these investigators could have been disclosed and that transparency would have facilitated additional trust and potential to verify and replicate the findings. T2 indicates other thresholds and sensitivity levels for the model as well, such as that the odds movement calculation only had to reach the 10% level for one out of the seven bookmakers being analyzed in order to be further considered.

At the inference layer of the project it is clear from the transparency disclosures what is being calculated (i.e. odds-movement) and how those calculations are then used in a simulation process to infer suspicious activity on the part of players. Moreover, additional information about the confidence of inferences is given by using statistical significance testing including Bonferroni correction to reduce the possibility of false positives. The final report lists suspicious players (anonymously) labeled according to their statistical significance at difference standard levels including $p < 0.05$, $p < 0.01$, $p < 0.001$, as well as listed for the Bonferroni corrected level. This allows us to see the degree of statistical confidence in the inference and judge the possibility of false positives in the final set of 15 players. Conspicuously missing from T2 or T3 however is a disclosure indicating error analysis. The closest we get is a note in the T3 Jupyter notebook stating, "In some simulations an additional player received an estimated likelihood just barely under 0.05. To be conservative we are not including that player among our totals." But this implies that there were multiple runs of the simulation procedure and that additional error analysis might have been tabulated, or even confidence intervals listed around each of the likelihood values reported.

In this case, the algorithmic transparency model was useful for identifying gaps in the information disclosures surrounding the methods used. This was a very solid project, done by seasoned reporters, and many of the dimensions of disclosure enumerated in the model were indeed covered. But there were some opportunities for additional disclosure around the data sample used, around the rationale for thresholds, and around the error bounds of the inferences. These criticisms elucidate opportunities to have bullet-proofed the disclosed methods even further. The implications for a broader public are limited since much of this criticism focuses on the lower levels (T2, T3) of more technical information disclosure. Mostly the story and its methods serve to spur additional investigation by sanctioned tennis investigators who have access to all of the necessary records in order to investigate suspicious players.

The journalists who worked on this story were thoughtful about not naming the players suggested by their methodology. Heidi Blake, the first author of the main article, works as the UK investigations editor for BuzzFeed News and is based in London, England. Since British law makes it considerably easier to sue for libel than in the US [33], one might speculate that BuzzFeed News thought the statistical evidence was not strong enough to stave off libel lawsuits in that jurisdiction. At the same time, the level of transparency in the project, particularly in T2 and T3 allowed for a team of three undergraduate students to ostensibly de-anonymize the results relatively quickly and easily.[12] The students re-scraped the odds information (though there was some uncertainty in identifying the seven used in the original story) with identities preserved, and then cross-referenced with the BuzzFeed data based on the other data fields available in order to associate a name with each of the 15 players identified in the original analysis. We can see this use of transparency information as undermining the intent of the journalists publishing the story, but we can also see it in a more constructive light. By stopping short of naming

[12]https://medium.com/@rkaplan/finding-the-tennis-suspects-c2d9f198c33d#.q1axxecwd.

names BuzzFeed news was able to avoid the risk of potential libel lawsuits. But by putting enough information and method out there so that others could replicate the work with their own non-anonymized data, this allowed a less legally averse outlet (students blogging about it) to publish the names. Presumable the same replication could be done by tennis investigators who would also be interested in the specific identities of players suspected of match fixing.

4 Discussion

In this chapter I have presented three cases in which an algorithmic transparency model informed both constructive and critical approaches towards media accountability. In the first case the model was used to guide the disclosure of editorial information about a news bot that might be useful to other journalists as well as to end-users. In the second case, the model suggested questions that were used to orient an investigation of the biases embedded in a news product tied to Google search rankings. It guided certain questions that could be directed towards the company, such as confirming the use of algorithms, and eliciting information about the input variables and data. And in the final case, the model identified gaps in the methodological transparency disclosed in a computationally informed piece of investigative journalism, suggesting opportunities that could have strengthened the disclosure from a replicability standpoint. This diversity in cases is important as it shows the versatility of the algorithmic transparency model employed and its ability to inform a range of activities that facilitate media accountability: of algorithmically-driven news products as well as algorithmically informed news investigations that result in a more traditional story. The model was found to be helpful in thinking through the different types of information that might be either voluntarily disclosed or elicited through interactive questioning or observation of an algorithm.

Still, there are a number of open questions and challenges in regards to algorithmic transparency and media accountability. One criticism that has been leveled at algorithmic transparency is that by disclosing too much information this would lead to a cognitive complexity challenge that overwhelms end-users who then don't know what to make of that information [19]. The BuzzFeed case addressed this issue by presenting information in a hierarchy of increasing detail, allowing end-users to choose the fidelity of transparency information that fit with their needs. The critique presented in this chapter suggests there might be even more nuance and detail that could be layered into these disclosures. But there is a key bit of additional work to do here. Research still needs to be done with end-users to see if such an approach is effective from their point of view. Such user research might report the rate of access or use of transparency information across different levels, or it might ask end-users explicitly what they found valuable from the disclosure at a particular level. It's not that user feedback should strictly dictate what is disclosed, but it may impact the interface and presentation of the disclosure. More generally, research also still needs to answer the question of the magnitude of end-user demand for algorithmic transparency, and the specific contexts and circumstances in which such disclosures do become important to a broad public.

Another criticism of increased algorithmic transparency is the cost of producing additional information disclosures. Such efforts should not be underestimated as they do create an additional time and resource burden on the creators of algorithmic systems, both in terms of the initial creation of that information but also in terms of maintenance. For one-off stories like the tennis investigation the ongoing cost is probably not that substantial. BuzzFeed would merely need to update the code or documentation if there is a major new revelation or if someone submits a bug fix (roughly parallel to the cost of posting a correction in traditional media). BuzzFeed needed to produce the code and comments in order to do the story anyway, so the additional cost of making it public is incremental. But for platforms or products, open sourcing code could lead to substantially more time and effort spent on maintaining that code base including "customer-service" related to answering questions from other users interested in adopting the code. Moreover, algorithmic systems are often dynamic and depending on how frequently the product changes this raises the question and challenge of keeping transparency disclosures up-to-date with the latest version of the product. If a product opts not to disclose detailed data, code, or methodology, the cost of producing adequate transparency documentation could be reasonable though. In one personal experience working with a data analytics startup on a transparency disclosure for a new product, it took roughly 5 h including interviewing an engineer along the lines of the algorithmic transparency model, and writing up the disclosure in accessible language that was then vetted by the product manager. Still, more work needs to be done, perhaps by workshopping the production of such disclosures with many more members of industry, so that the bottlenecks and resources needed in order to create and publish meaningful transparency information can be better understood.

Transparency information was made available on the Github platform in both the news bot as well as the tennis investigation case studies. Github is a popular service that allows for putting code and documentation into online repositories that can be kept private or made public for open source projects. This has several advantages including the ability to version manage the disclosure, track any edits or changes, fork into new projects, track social interest through repository starring activity, and render commingled code and descriptive notes in Jupyter Notebook format. At the same time there are constraints to Github, such as strict file size limits that make the platform unsuitable for depositing large data files. For open and extensible projects like news bots, or investigative methodologies, platforms like Github serve well at the medium to low-levels of transparency. But presentation of transparency information that way might make less sense in the case of something like the Google infoboxes. In that case a written disclosure in the form of a blog post might have been sufficient since code-level detail is not strictly necessary (or warranted) for a computational product. Products surface different types of concerns around transparency, such as the loss of trade secrecy, or the weakening of a system with respect to manipulation or gaming. But while Google may be averse to showing the code and the exact methods in which candidate quotes are integrated into their search engine, there are other factors (e.g. variables of interest, rationale for ordering decisions) that could be explained in a written disclosure so that end users have a higher degree of reliance on the information they receive. An open question for research is to determine the most appropriate vehicle, format, or platform to publish transparency information given the intended end-users and their goals. In some cases it may make sense to integrate transparency information directly into the user interface, but in others having that information linked on a repository, or published in a blog post or whitepaper may be more appropriate.

In closing, I would reiterate that there is still much work to do in studying algorithmic transparency. This chapter has shown some of the nuances in transparency around product-oriented computation versus editorial-oriented computation. Additional research should examine such differences and connect these to concerns over issues of proprietary trade secrecy or concerns over manipulation. The model applied in this chapter should be seen as a starting point that was helpful for enumerating different types of information that might be disclosed about algorithms, both in the context of using algorithms to create media, as well as in the context of investigating and critiquing others' use of algorithms in the media.

References

1. Graefe, A.: Guide to Automated Journalism. Tow Center for Digital Journalism, New York, NY (2016)
2. Lokot, T., Diakopoulos, N.: News bots: automating news and information dissemination on Twitter. Digit. J. **4**, 682–699 (2016)
3. Spangher, A.: Building the Next New York Times Recommendation Engine. New York Times, New York, NY (2015)
4. Wang, S., Han, E.-H. (Sam), Rush, A.M.: Headliner: an integrated headline suggestion system. In: Computation C Journalism Symposium, Palo Alto, CA (2016)
5. Wang, S., Han, E.-H.: BreakFast: analyzing celerity of news. In: International Conference on Machine Learning and Applications (ICMLA), Miami, FL (2015)
6. Magnusson, M., Finnäs, J., Wallentin, L.: Finding the news lead in the data haystack: automated local data journalism using crime data. In: Computation + Journalism Symposium, Palo Alto, CA (2016)
7. Thurman, N., Schifferes, S., Fletcher, R., et al.: Giving computers a nose for news. Digit. J. **4**(7), 743–744 (2016). doi:10.1080/21670811.2016.1149436
8. Park, D.G., Sachar, S., Diakopoulos, N., Elmqvist, N.: Supporting comment moderators in identifying high quality online news comments. In: Proceedings of the Conference on Human Factors in Computing Systems (CHI) (2016)
9. Gillespie, T.: The relevance of algorithms. In: Media Technologies: Essays on Communication, Materiality, and Society. The MIT Press, Cambridge, MA (2014)
10. Pasquale, F.: The Black Box Society: The Secret Algorithms That Control Money and Information. Harvard University Press, Cambridge, MA (2015)
11. Diakopoulos, N.: Algorithmic accountability: journalistic investigation of computational power structures. Digit. J. **3**, 398–415 (2015)
12. Ward, S.J.A.: Radical Media Ethics: A Global Approach. Wiley-Blackwell, Malden, MA (2015)

13. Fink, K., Schudson, M.: The rise of contextual journalism, 1950s–2000s. Journalism. **15**, 3–20 (2014)
14. McBride, K., Rosenstiel, T.: The new ethics of journalism: principles for the 21st century. In: The New Ethics of Journalism: Principles for the 21st Century. Sage, Thousand Oaks, CA (2013)
15. Friedman, B., Nissenbaum, H.: Bias in computer systems. ACM Trans. Inf. Syst. **14**, 330–347 (1996)
16. Fengler, S., Russ-Mohl, S.: The (behavioral) economics of media accountability. In: Fengler, S., Eberwein, T., Mazzoleni, G., Porlezza, C. (eds.) Journalists and Media Accountability: An International Study of News People in the Digital Age, pp. 213–230. Peter Lang, New York (2014)
17. Deuze, M.: What is journalism?: professional identity and ideology of journalists reconsidered. Journalism. **6**, 442–464 (2005). doi:10.1177/1464884905056815
18. Silverman, C.: Corrections and ethics. In: McBride, K., Rosenstiel, T. (eds.) The New Ethics of Journalism—Principles for the 21st Century, pp. 151–161. Sage, Thousand Oaks, CA (2013)
19. Diakopoulos, N., Koliska, M.: Algorithmic transparency in the news media. Digit. J. (2016). http://nca.tandfonline.com/doi/abs/10.1080/21670811.2016.1208053
20. Lofland, J., Lofland, L.: Analyzing Social Settings: A Guide to Qualitative Observation and Analysis, 3rd edn. Wadsworth Publishing Company, Belmont, CA (1995)
21. Burrell, J.: How the machine "thinks": understanding opacity in machine learning algorithms. Big Data Soc. **3**(1), 1–12 (2016)
22. Fung, A., Graham, M., Weil, D.: Full Disclosure: The Perils and Promise of Transparency. Cambridge University Press, Cambridge (2009)
23. Stark, J., Diakopoulos, N.: Towards editorial transparency in computational journalism. In: Computation + Journalism Symposium, (2016)
24. Trielli, D., Mussende, S., Stark, J., Diakopoulos, N.: How the Google Issue Guide on Candidates is Biased. Slate (2016)
25. Napoli, P.M., Caplan, R.: When media companies insist they're not media companies and why it matters for communications policy. In: Telecommunications Policy Research Conference (2016)
26. Epstein, R., Robertson, R.E.: The search engine manipulation effect (SEME) and its possible impact on the outcomes of elections. Proc. Natl. Acad. Sci. U.S.A. **112**, E4512–E4521 (2015)
27. Purcell, K., Brenner, J., Rainie, L.: Search Engine Use 2012. Pew Internet and American Life Project, Washington, DC (2012)
28. Agichtein, E., Brill, E., Dumais, S., Ragno, R.: Learning user interaction models for predicting web search result preferences. In: Proceedings of SIGIR 2006 (2006)
29. Meyer, P.: Precision Journalism: A Reporter's Introduction to Social Science Methods, 4th edn. Rowman & Littlefield Publishers, Lanham, MD (2002)
30. Goschowski Jones, R., Ornstein, C.: Matching Industry Payments to Medicare Prescribing Patterns: An Analysis (2016). https://static.propublica.org/projects/d4d/20160317-matching-industry-payments.pdf?22
31. Blake, H., Templon, J.: The Tennis Racket. BuzzFeed News (2016). https://www.buzzfeed.com/heidiblake/the-tennis-racket
32. Templon, J.: How BuzzFeed News Used Betting Data to Investigate Match-Fixing in Tennis. BuzzFeed News (2016). https://www.buzzfeed.com/johntemplon/how-we-used-data-to-investigate-match-fixing-in-tennis
33. Shapiro, A.: On Libel and the Law, U.S. and U.K. Go Separate Ways. National Public Radio (NPR) (2015)

The Princeton Web Transparency and Accountability Project

Arvind Narayanan and Dillon Reisman

Abstract When you browse the web, hidden "third parties" collect a large amount of data about your behavior. This data feeds algorithms to target ads to you, tailor your news recommendations, and sometimes vary prices of online products. The network of trackers comprises hundreds of entities, but consumers have little awareness of its pervasiveness and sophistication. This chapter discusses the findings and experiences of the Princeton Web Transparency Project (https://webtap.princeton.edu/), which continually monitors the web to uncover what user data companies collect, how they collect it, and what they do with it. We do this via a largely automated monthly "census" of the top 1 million websites, in effect "tracking the trackers". Our tools and findings have proven useful to regulators and investigatory journalists, and have led to greater public awareness, the cessation of some privacy-infringing practices, and the creation of new consumer privacy tools. But the work raises many new questions. For example, should we hold websites accountable for the privacy breaches caused by third parties? The chapter concludes with a discussion of such tricky issues and makes recommendations for public policy and regulation of privacy.

1 Introduction

In 1966, Marvin Minsky, a pioneer of artificial intelligence, hired a freshman undergraduate for a summer to solve 'the vision problem', which was to connect a TV camera to a computer, and get the machine to describe what it sees [1]. Today this anecdote is amusing, but only because we understand with the benefit of hindsight that many, or even most things that are easy for people are extraordinarily hard for computers. So the research field of AI barked up a few wrong trees, and even had a couple of so-called "AI winters" before finally finding its footing.

A. Narayanan (✉) • D. Reisman
Princeton University, Princeton, NJ, USA
e-mail: arvindn@cs.princeton.edu; dreisman@princeton.edu

T. Cerquitelli et al. (eds.), *Transparent Data Mining for Big and Small Data*,
Studies in Big Data 32, DOI 10.1007/978-3-319-54024-5_3

Meanwhile, automated decision-making had long been recognized by industry as being commercially valuable, with applications ranging from medical diagnosis to evaluating loan applications. This field too had some missteps. In the 1980s, billions of dollars were invested in building "expert systems", which involved laboriously creating databases of facts and inference rules from which machines could supposedly learn to reason like experts. An expert system for medical diagnosis, for example, would rely on physicians to codify their decision-making into something resembling a very large decision tree. To make a diagnosis, facts and observations about the patient would be fed into this set of rules. While such expert systems were somewhat useful, they ultimately they did not live up to their promise [2].

Instead, what has enabled AI to make striking and sustained progress is "machine learning". Rather than represent human knowledge through symbolic techniques, as is done in expert systems, machine learning works by mining human data through statistical means—data that is now largely available thanks to the "Big Data" revolution. Machine learning has become a Silicon Valley mantra and has been embraced in all sorts of applications. Most famously, machine learning is a big part of online personalized advertising—as data scientist Jeff Hammerbacher said, "The best minds of our generation are thinking about how to make people click on ads." [3]

But machine learning has also made its way into more important applications, like medical diagnosis and the determination of creditworthiness. It's even being employed in "predictive policing" and the prediction of criminal recidivism, where it has life-or-death consequences [4]. Machine learning has also proved extremely adept at the computer vision problems that Minsky was interested in 50 years ago, as well as related domains such as natural-language processing.

So machine intelligence and automated decision-making today increasingly rely on machine learning and big data. This brings great benefits ranging from movie recommendations on Netflix to self-driving cars. But it has three worrying consequences. The first is privacy. Many machine learning systems feed on people's personal data—either data about them or data created by them. It can be hard to anticipate when and how a piece of personal data will be useful to make decisions, which has led to a culture of "collect data first, ask questions later." This culture has spurred the development of an online surveillance infrastructure that tracks, stores, and profiles everything we do online.

The second consequence to society is that the outputs of machine learning reflect human biases and prejudices. One might have naively assumed that AI and machines in the course of automated decision-making would somehow be mathematically pure and perfect, and would make unbiased decisions. Instead, we're finding that because the data used to train machine learning models comes from humans, machines essentially inherit our biases and prejudices [5].

The third concern over our use of machine learning is what you might call the inscrutability of AI. It may have once been correct to think of automated decision-making systems as some sort of decision tree, or applying a set of rules. That's simply not how these systems work anymore. When a machine learning system is trained on a corpus of billions of data points to make medical diagnoses, or to serve ads online, it is impossible to express to a patient or a user, or even to the creator of the system, the reason why the machine decided as it did. When we put these complex, inscrutable decision-making systems in a position of power over people, the result can be Kafkaesque [6].

These three concerns have led to much-needed public and scholarly debate. In addition, the inscrutability of these systems has necessitated a new kind of empirical research that can peek into the black boxes of algorithmic systems and figure out what's going on. This new research field, which we call "data and algorithmic transparency", seeks to address questions about the data collection and use that happens around us everyday, questions such as:

- Are my smart devices in my home surreptitiously recording audio?
- Does my web search history allow inferring intimate details, even ones I've not explicitly searched for?
- Is the algorithm that decides my loan applications biased?
- Do I see different prices online based on my browsing and purchase history?
- Are there dangerous instabilities or feedback loops in algorithmic systems ranging from finance to road traffic prediction?

A combination of skills from a variety of areas of computer science is called for if we are going to learn what is inside the black boxes. We need to build systems to support large-scale and automated measurements, and modify devices to record and reverse engineer network traffic. We need expertise to simulate, model, and probe decision-making systems. And we need to reach out to the public to collect data on the behavior of these algorithmic systems in the wild.

But more than just computer scientists, this effort is bringing together a new interdisciplinary community of empirical researchers, journalists, and ethical scholars, with new conferences and workshops focusing on transparency research, such as the workshop on Data and Algorithmic Transparency (http://datworkshop.org/). In an ideal world, one would imagine that companies involved in building data-driven algorithms would be perfectly forthcoming about what personal data they're using and how these systems work behind the scenes, but unfortunately that is not the world we live in today. Even if companies and governments were forthcoming, analyzing and understanding the societal impact of these systems will always require empirical research and scholarly debate.

2 The Princeton Web Transparency and Accountability Project

We started the "Princeton Web Transparency and Accountability Project" (WebTAP) to focus our study on the web, a rich source of data that feeds into decision-making systems, and a prominent arena where their effects can be seen. A major output of WebTAP is the "Princeton Web Census" which is an automated study of privacy across 1 million websites which we conduct every month.

Our work so far has focused on monitoring and reverse-engineering web tracking. Given the inadequacy of laws and rules that govern web tracking, we believe that external oversight of the online tracking ecosystem is sorely needed. We're interested in questions like, "Which companies track users? What technologies are they using? What user data is being collected? How is that data being shared and used?"

2.1 The Perils of Web Tracking

What are the potential harms of web tracking? The first is simply that the erosion of privacy affects our intellectual freedom. Research shows when people know they're being tracked and surveilled, they change their behavior [7]. Many of today's civil liberties—say, marriage equality—were stigmatized only a few decades ago. The reason it became possible to discuss such issues and try to change our norms and rules is because we had the freedom to talk to each other privately and to find like-minded people. As we move to a digital world, is ever-present tracking hindering those abilities and freedoms?

A second worrisome effect of web tracking is the personalization and potential discrimination that results from the use of our personal data in these algorithmic systems. Some online retailers have experimented with price discrimination, showing different prices to different visitors based on personal factors [8]. Without oversight, we lose the ability to limit practices that would be censured in the offline world.

There are many other domains which might be impacted through algorithmic personalization. Privacy scholar Ryan Calo argues that personalized digital advertising can be construed as a form of market manipulation [9]. There are also impacts in the political sphere, with consequences for the health of democracy. What happens when political campaigns start to use personal data in order to microtarget the messages that they send to voters, to tell slightly different stories to different people online through targeted advertisements?

Finally, the lack of transparency in online tracking is problematic in and of itself. There's no public input into the decision-making that happens in these systems, leading to unaccountable and opaque processes that have real consequences. We need to close that transparency gap. This is a large part of what motivates the Princeton WebTAP project.

2.2 How Web Tracking Works

In WebTAP, we mostly study "third party online tracking". When you go to nytimes.com, the New York Times knows you've visited and which article you're reading—in this case, the New York Times is a "first party". Because you choose to visit a first party, we are not particularly concerned about what the first party knows from your visit. Third party online tracking, however, happens when entities other than the one you're currently visiting compile profiles of your browsing history without your consent or knowledge [10]. While most third parties are invisible, visible page elements such as Facebook Like buttons, embedded Twitter feeds, and a variety of other commercial widgets are also modes of third party tracking. One study a few years ago showed that, on the average top 50 website, there are 64 independent tracking mechanisms [11]! That is consistent with our own findings—in fact, that number has only grown over time [12].

Web cookies are the most widely used mechanism for tracking on the web by first and third parties, a fact many users are already familiar with. What is less well-known is that, increasingly, websites and trackers are turning to techniques like browser fingerprinting—techniques that are sneakier, harder for users to protect themselves from, and which work without necessarily leaving any trace on your computer. The *Beauty and the Beast* project (https://amiunique.org/) and the older *Panopticlick* project (https://panopticlick.eff.org/) offer a demonstration of fingerprinting using a variety of attributes from your own web browser, such as the type and version number of the browser, the list of fonts that you have installed, the list of browser plugins you have installed, and more [13, 14]. Third parties can use such fingerprints to uniquely identify your device, no cookies required.

You might think that this tracking is anonymous, since your real name is not attached to it. The online advertising industry has repeatedly sought to reassure consumers this way. But this is a false promise. Many third parties do know your real identity. For example, when Facebook acts as a third party tracker they can know your identity as long as you've created a Facebook account and are logged in—and perhaps even if you aren't logged in [15]. Third parties with whom you don't have an account have many ways of inferring a user's real identity as well. Sometimes all that is needed are bugs and poor web development, resulting in personal identifiers "leaking" from first parties to third parties [16, 17]. It is also possible for a tracker to de-anonymize a user by algorithmically exploiting the statistical similarity between their browsing history and their social media profile, as demonstrated in a recent collaboration between Stanford researchers and WebTAP [18].

Even ignoring all this, web tracking is not anonymous. "Anonymous" means that an individual's activities (say, visits to different websites) cannot be linked together, but such linking is the entire point of third-party web tracking. "Pseudonymous" means that those activities can be linked together, even if the real-world identity behind them is unknown. As Barocas and Nissenbaum have argued, most potentially undesirable effects of tracking happen even if you're being tracked under a pseudonymous identity instead of your real identity [19]. The potential biases and discrimination that we discussed—targeted political messaging, price discrimination, market manipulation—can still happen online, as long as advertisers have some way to communicate to you.

2.3 The Open Web

Third party tracking is rampant in part because the web is built on open technology, and so the barriers to entry are low. For example, once there is one third party on a page, that third party has the ability to turn around and invite any number of other third parties to the first party webpage. Web technology standards and specifications are the primary regulator of privacy practices, and these tend to be permissive. As we'll see in Sect. 4.1, new HTML features introduced by standards bodies have repeatedly been repurposed for infringing privacy in devious and inventive ways. In contrast, in a closed platform such as Apple's app store, Apple exercises significant control over what behaviors are acceptable [20].

But by the same token, the web's openness is also good news for users and browser privacy tools. The browser ultimately acts on behalf of the user, and gives you—through extensions—an extraordinary degree of control over its behavior. This enables powerful tracker-blocking extensions to exist, which we'll revisit in Sect. 4.4. It also allows extensions to customize web pages in various other interesting ways ranging from making colors easier on the eyes to blocking ads.

A recent WebTAP paper demonstrates how powerful this capability can be. We set out to analyze the so-called ad-blocking wars and to predict how ad-blocking technology might evolve [21]. Web publishers and other news sites are unhappy about the increasing popularity of ad-blockers among users. One response has been the deployment of ad-blocker blockers, to prevent users from viewing content unless the ad blocker is disabled. We argue that this strategy could not succeed in the long run—a determined user will always be able to block ads. Ad blocking extensions can modify the browser in powerful ways, including hiding their own existence from the prying JavaScript code deployed by websites. Ad blockers will likely ultimately succeed at this because browser extension code executes at a higher "privilege level" than website code.

Other websites, most notably Facebook, are trying to make their ads indistinguishable from regular posts, thus making it harder for ad-blockers to block ads without also blocking real user content. This is again a dubious strategy in the long run. Due to the enforcement of deceptive advertising rules by the US Federal Trade Commission, human users have to be able to tell ads and regular content apart. Ad industry self-regulatory programs such as AdChoices also have the same effect. We created a proof-of-concept extension to show that, if humans are able to distinguish ads and content, then automated methods could also be able to distinguish them, by making use of the same signals that humans would be looking at, such as the text "Sponsored" accompanying an advertisement [22, 23].

The broader lesson is that open technologies shift the balance of power to the technologically savvy. Another consequence of this principle is that the web's openness is good news for us as researchers. It makes the technical problem of automated oversight through web privacy measurement much easier. Let us give you a small but surprising example to illustrate this point.

A popular service provider on the web, called Optimizely, helps websites do "A/B testing." It is the practice of experimenting with different variations of a part of the website to evaluate how positively users respond to the variations. For instance, Optimizely allows publishers like New York Times to test two different versions of a headline by showing one version to 50% of visitors, and another version to the other 50%, so that they can evaluate which headline more users clicked on. Because of the open nature of the web, the code Optimizely uses to implement the experiments is actually exposed on an invisible part of the webpage—any visitor on the page who knows how to peek behind the scenes of the web browser could see every experiment that the website was doing on its users!

So we did a little digging. As part of the WebTAP project, we visited a number of different sites, and grabbed all of the Optimizely code and experimental data from sites that used Optimizely. You can find the full details of our study in our blog post, but there were some interesting instances of A/B testing that stood out [24]. Many news publishers experiment with the headlines they feature, in a questionable way. For example, a headline such as "Turkey's Prime Minister quits in Rift with President" might appear to a different user as "Premier to Quit Amid Turkey's Authoritarian Turn." You can see clearly that the implication of the headline is different in the two cases. We also found that the website of a popular fitness tracker targets users that originate from a small list of hard-coded IP addresses labeled "IP addresses spending more than $1000." While amusing, this can also be seen as somewhat disturbing.

3 WebTAP's Main Engine: OpenWPM

Back in 2013, when we started WebTAP, we found that there had been over 20 studies that had used automated web browsers for studying some aspect of privacy or security, or online data-driven bias and discrimination. We found that many of those studies had devoted a lot of time to engineering similar solutions, encountering the same problems and obstacles. We were motivated to solve this problem and share it with the community, which led to our development of OpenWPM—Open Web Privacy Measurement.[1] It builds on ideas and techniques from prior projects, especially FourthParty [10] and FPDetective [25]. Today it is a mature open source project that has a number of users and researchers using it for a variety of studies on online tracking.

[1] OpenWPM is available for download at https://github.com/citp/OpenWPM.

3.1 Problems Solved by OpenWPM

OpenWPM solves a number of problems that researchers face when they want to do web privacy measurements. First, you want your automated platform to behave like a realistic user. Many researchers have used a stripped-down browser, such as PhantomJS, that does not fully replicate the human browsing experience [26]. While this might be okay for some experiments, it won't reproduce what happens when a real user browses the web. To solve this problem the OpenWPM platform uses a full version of the popular Firefox web browser.[2]

OpenWPM also allows simulating users with different demographics or interest profiles by building up a specified history of activity in the automated browser. For example, before beginning an experiment a researcher can have OpenWPM visit sites that are popular among men or among women, depending on the profile they are trying to mimic.

The second problem that OpenWPM solves is that it collects all data that might be relevant to privacy and security concerns from the automated web browser. This includes all of the network requests the automated browser makes, including cookies, but also information that's relevant to browser fingerprinting. OpenWPM has the ability to log every single JavaScript call to a browser JavaScript API that is made by a script on the page, and the identity of the third party or first party that made the call. We have found this feature very useful in our own experiments.

This set of information is collected from three different vantage points. A browser extension collects the JavaScript calls as they are made in the browser. A "man in the middle" proxy intercepts network requests made between the browser and the website it visits. Lastly, OpenWPM collects any information that is stored on the computer's disk, such as flash cookies. OpenWPM unifies these views of the data and stores them in a single database that provides an easy interface for analysis.

OpenWPM also makes it easier for researchers when it comes time for data analysis. We provide tools that can aid researchers in answering simple questions like, "What are all of the third parties that were ever seen on a particular first party site?" More complex analyses can be easily built off of those building blocks as well.

The stability of the tools that are used for web measurement has also been a problem for past researchers. OpenWPM uses a tool called Selenium to automate the actions of the web browser, as done in many prior web privacy measurement experiments [27]. While Selenium makes issuing commands to Firefox very simple, it was meant for testing single websites. In our experiments we visit as many as one million websites every month. Selenium, we've discovered, was not made to scale to that many websites, and without "babysitting" by the researcher, will often crash in the middle of an experiment. OpenWPM solves the stability issue by recovering from crashes in Selenium while preserving the state of the browser.

[2]It is possible to use another browser, such as Chrome, in OpenWPM, but in our experiments so far we've used Firefox.

3.2 *OpenWPM's Advanced Features*

We're also working on a goal we call "one-click reproducibility," which would give researchers the ability to reproduce the results of experiments done by others. OpenWPM issues commands that could be stored in a standard format and replayed later. A package containing these commands along with analysis scripts used to evaluate the experiments and its results could be shared to enable collaboration among researchers and verification of experimental results. One possible use-case for one-click reproducibility can be found in the world of regulation—if a regulator conducts a study involving a complex series of commands across a set of websites, then they may want the ability to share the exact tools and procedure they used with other parties concerned with the regulation. Beyond specific use-cases, we hope that this will help scientific replication of web privacy studies.

OpenWPM also has a number of advanced features that are specifically useful for privacy studies. OpenWPM can automatically detect which cookies are unique tracking cookies and hence privacy-relevant (as opposed to cookies setting website language, or timezone). OpenWPM also has a limited ability to automatically login to websites using "federated login mechanisms". Federated login mechanisms, like Facebook Connect or Google's Single Sign-On (SSO), are tools that, when added to a website, allow users of that website to use their Facebook or Google account to sign-in. Since many websites leverage federated login mechanisms, our ability to use it to automatically login can be quite useful since certain privacy impacting behaviors are only triggered when a user is logged in.

OpenWPM also has a limited ability to extract content from webpages. For example, if a researcher wants to measure how search results vary between users, there is a way for them to specify how to extract the search results from a webpage. While the process is not yet entirely automated, it only requires a few lines of code. OpenWPM stores such extracted content for later analysis.

4 WebTAP's Findings

Frequently, we've found, a privacy researcher publishes a study that finds a questionable privacy-infringing practice. This puts public scrutiny on the third parties or websites concerned, leading to a temporary cessation of the practice, only for the episode to be forgotten a month later, at which time the privacy-infringing behavior creeps back into use.

WebTAP avoids these problems of "one-off" privacy studies because our studies are *longitudinal*. Our analyses are automated, which allows us to continually monitoring questionable practices on the web and keep pressure on trackers to avoid those practices.

4.1 Detecting and Measuring Novel Methods of Fingerprinting

Through our crawls of the web we've found that third-party tracking technologies evolve rapidly. In particular, we've tracked the evolution of browser fingerprinting techniques. Most or all of the information provided by your browser to websites and third parties can be used to fingerprint your device and to track you. As long as a piece of information is at least somewhat stable over time, it can be leveraged in a fingerprint. This can include things as innocuous as the size of your browser window, the version number of your browser, or even the set of fonts installed on your computer.

Imagine that an online tracker derives a fingerprint for you that consists of, say, 43 different features. After the tracker first sees you, however, you install two new fonts on your computer. When the tracker sees you again, you'll have a fingerprint with 45 features, of which 43 will match the previous observation. Much like observations of actual fingerprints, two such observations that differ slightly can still be linked through statistical means. In the industry, these fingerprints are often called "statistical IDs."

In a 2014 study, we collaborated with researchers at KU Leuven to study a technique called "canvas fingerprinting" [28]. The Canvas API, recently added to browsers in the HTML5 standard, gives scripts a simpler interface for drawing images on webpages. Scripts can also use the canvas to draw an image that is invisible to the user. Once the image is drawn on the webpage, the script can read the image data back pixel by pixel. It turns out that the precise pixel-wise representation of the image, such as the one seen in Fig. 1, will vary between different devices based on unique features of the rendering software that is used for displaying images on the screen. The Canvas API is one of the sneakier ways that a seemingly benign interface provided by the browser can contribute to a unique fingerprint.

Canvas fingerprinting had been first proposed in a security paper in 2012 [29]. Some time later a developer implemented an open-source library that included canvas fingerprinting, and a few obscure sites experimented with it [30]. But by the time of our study in early 2014, a mere year and a half after it had first been proposed, several third parties, including a major online third party called AddThis, had employed canvas fingerprinting—of the top 100,000 websites included in

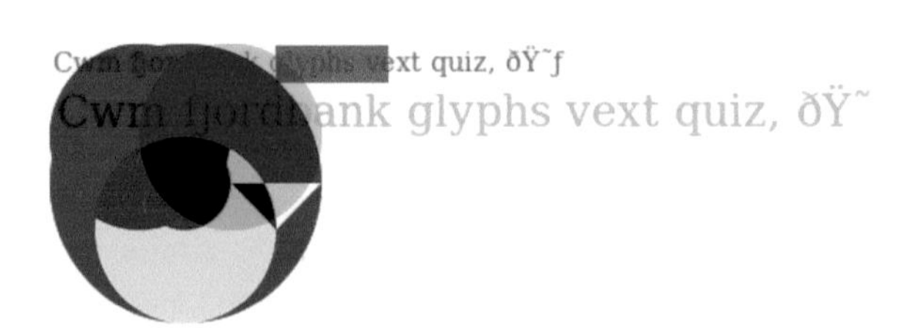

Fig. 1 The image, invisible to the user, that is drawn on the web page by one of the canvas fingerprinting scripts that we detected. The image is converted to a string of data, constituting the fingerprint

our study, over 5% had third party scripts that employed the technique. This is one example of how obscure tracking techniques can quickly go from academic curiosities to becoming mainstream on the web. This discovery was only possible through the use of automated measurement. This quickly led to improvements in browser privacy tools as well as a public and press backlash that led AddThis to drop the technique.

This study was done in 2014, and didn't use OpenWPM, as it wasn't mature yet. In 2016, we began WebTAP's monthly 1-million-site measurements using OpenWPM. We've found that virtually every HTML5 API is being abused to fingerprint devices. This includes the AudioContext API for audio processing, the Battery Status API [31], and the WebRTC API for peer-to-peer real-time communication. We were able to catch these techniques early in their lifecycles, when they were deployed on relatively small numbers of sites [32]. Our ongoing findings have led to debate and discussion in the web standards community on how to design these APIs in a way that resists the ability to utilize them for fingerprinting [33].

4.2 The "Collateral Damage" of Web Tracking

WebTAP has shed light on how the online tracking infrastructure built by web companies for commercial purposes can be repurposed for government surveillance. The Snowden leaks revealed that the NSA has in fact been reading tracking cookies sent over networks to enable their own user tracking [34]. We wanted to study and quantify just how effective this could be.

The answer isn't obvious. First, the technique might not work because any given tracker might appear on only a small fraction of web pages—it is unclear to what extent the NSA or another surveillance agency might be able to put together a complete picture of any given person's web browsing traffic using just these cookies. Second, cookies are pseudonymous, as we discussed earlier. Even though Facebook, for example, might know the real-world identity associated with each cookie, it will not necessarily send that information across the network where the NSA can read it. There are various other complications: for example, it is not clear what portion of a typical user's traffic might pass through a vantage point on the internet that the NSA can observe.

Using OpenWPM, we simulated a typical user browsing the web and analyzed which trackers were embedded on which websites [35]. Even if a tracker is not embedded on a significant number of websites, if two different trackers are embedded on the same page, an eavesdropper on the network can infer that the two tracker's distinct IDs belong to the same user. With enough trackers embedded on enough websites, it is possible to transitively link all of the tracking IDs for a single user using the procedure illustrated in Fig. 2.

Using this notion of transitive cookie linking, as well as using geographical measurements of where websites and their servers are located, we were able to

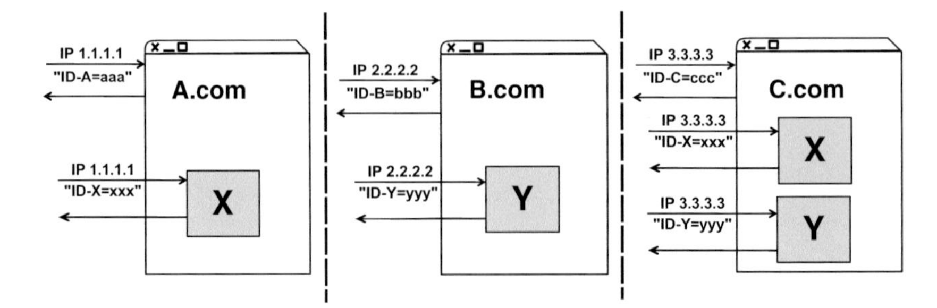

Fig. 2 How cookie linking works: An eavesdropper observes identifiers ID-X and ID-Y on two separate page loads. On a third page load, both identifiers are seen simultaneously, allowing the eavesdropper to associate ID-X and ID-Y as belonging to the same user

Fig. 3 The email address of a logged-in New York Times user is displayed on the top banner of the website (highlighted in *red*). Because the New York Times does not deliver their webpage over an encrypted connection (as of September 2016), an eavesdropper can use the email address to assign a real-world identity to a pseudonymous web browser

show that an eavesdropper like the NSA will be able to reconstruct 60–70% of a user's browsing history. Since a number of sites have inadequate use of encryption on their websites, we were also able to show that such an adversary will very likely be able to attach real-world identities to these browsing histories for typical users. If a website fails to encrypt a page containing personally identifiable information, like a name or email address as seen on the New York Times website in Fig. 3, we found that it was likely that eventually a real identity would be leaked across the network along with the tracking cookies.

This finding highlights the fact that concerns about online tracking go beyond its commercial applications, and have consequences for civil society. It also underscores the importance of deploying encryption on the web—HTTPS is not just for protecting credit card numbers and other security-sensitive information, but also for protecting the privacy of the trails we leave online, which might otherwise be exploited by any number of actors.

Unfortunately, in the same theme of "collateral damage," our research has also shown that third parties on the web impede the adoption of HTTPS [32]. Our measurements reveal that many players in the online tracking ecosystem do not provide an encrypted version of their services. The web's security policies dictate, with good reason, that for a web page to deploy HTTPS, all the third-party scripts on the page also be served encrypted. About a quarter of unencrypted sites today would face problems transitioning to HTTPS due to third parties.

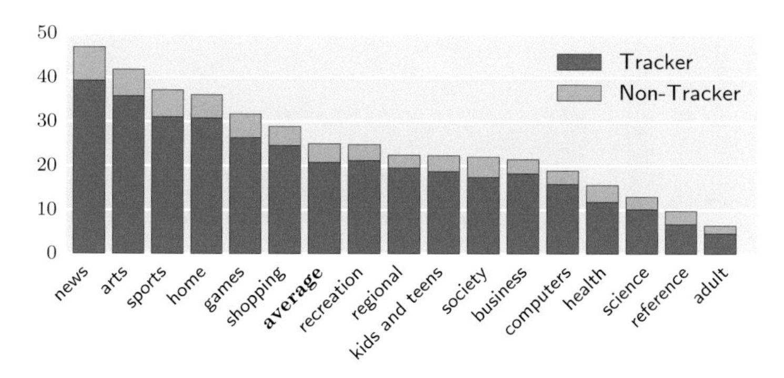

Fig. 4 The prevalence of third-parties by Alexa site category, split between tracking and non-tracking third parties

4.3 *The Economic Forces Behind Tracking*

WebTAP's findings are in line with the intuition that the need for ad revenue, especially among news publishers, is a big driver behind the prevalence of tracking [32]. Breaking down the number of trackers by website category, available in Fig. 4, we see that news sites have the most embedded trackers. Interestingly, adult sites seem to have the least. Perhaps adult websites are more concerned with user privacy, or perhaps fewer advertisers are interested in advertising on adult sites; we cannot know for sure from the data alone.

We've also found that the web ecosystem has experienced some consolidation among tracking companies. There is certainly a long tail of online tracking—in our study of one million websites, we found that there are over 80,000 third parties that are potentially in the business of tracking. Most of these, however, are only found on a small number of first-party sites. In fact, there are relatively few third parties, around 120, that are found on more than 1% of sites. And a mere six companies have trackers on more than 10% of websites, as shown in Fig. 5 [32].

Arguably, this is good for privacy, because the more popular third parties are larger companies like Google or Facebook that are more consumer-facing. When such companies get caught in a privacy misstep, it is more likely that the resulting negative public reaction will have consequences for them. These bigger players also get more scrutiny from regulatory agencies like the Federal Trade Commission and are, perhaps, more willing to police themselves. On the other hand, there's also an argument that the large concentrations of online tracking data that result from economic consolidation are bad for privacy.

Even ignoring economic consolidation, there are several ways in which different tracking databases get linked to or merged with each other, and these are unambiguously bad for privacy. Because of a process called "cookie syncing," individual trackers gain a greater ability to see more of your online behavior. Cookie syncing allows trackers to link their identifying cookies to other companies' cookies, giving a tracker's customers greater insights into a user's activity from more sources. In our

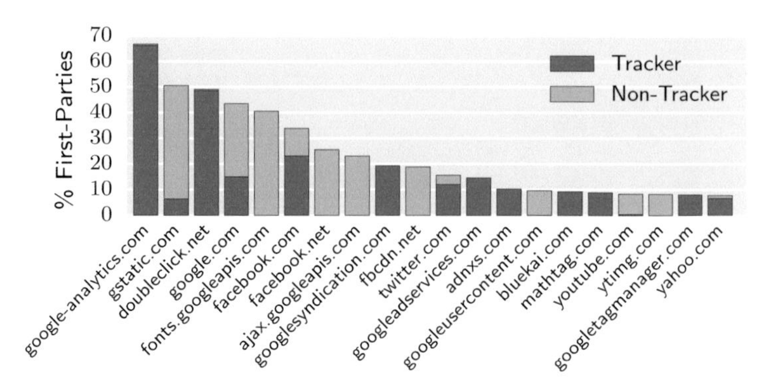

Fig. 5 The most prevalent third-parties by domain, split between when the third party was identified as tracking vs. non-tracking

census, we found that 85 of the top 100 most common third parties sync their cookies with at least one other party. Google was the most prolific cookie-syncer—there are *118* other third parties with which it shares cookies or which share cookies with it.

Researchers and journalists have documented several other trends towards consolidation of tracking databases. First, the *internal* privacy walls at tech companies are coming down [36–38]. "Cross-device tracking" links your different devices with each other—sometimes by invasive methods such as playing an ultrasound audio code on a web page and listening to it from your smartphone [39, 40]. "Onboarding" by firms including Facebook links your (physical) shopping records with your online activity [41]. The euphemistic "header enrichment" by Internet Service Providers adds the equivalent of a cookie to your web or mobile traffic—a single cookie for all your activities rather than a tracker-specific identifier [42].

4.4 The Impact of Web Privacy Measurement

We've found repeatedly that merely measuring online tracking has a positive impact on web privacy, due to increased public awareness and companies' desire to avoid the spotlight on privacy issues. This was a surprise for us—traditionally in computer security research, measurement is at best a first step to try and understand the scope of the problem before you start the process of devising solutions.

Measurement seems to mitigate the "information asymmetry" that exists in online tracking. That is, web trackers know a lot more about the technologies and the type of data that's being collected than consumers do. In fact, publishers are often in the dark about the extent of third party tracking on their own websites. In response to our study, many of the websites where canvas fingerprinting had been found responded to press inquiries to say that they were entirely unaware of the practice. Measurement seems to fix this information asymmetry for both users and website owners alike, making for a more informed public debate on these issues and bringing data to the table for better informed policy making.

Today the main way users can protect themselves against online tracking is by installing browser privacy tools such as Ghostery, AdBlock Plus, or uBlock Origin. Measurement research helps improve these tools—sometimes by finding entirely new categories of tracking and sometimes by finding new trackers employing known types of tracking. The block lists used by these tools were compiled by a laborious manual process, but automated methods allow us to find new trackers quickly and efficiently. The data released by our project finds its way into these tools [43].

Even so, today's tracker-blocking tools have important limitations: they block many benign URLs and break the functionality of a significant fraction of sites. In ongoing research, we are looking at the possibility that these tools can be built in a radically different manner: using machine learning to automatically learn the difference between tracking and non-tracking *behavior* instead of *actors*.[3] The machine-learning classifier would be trained on our web-scale census datasets, and the browser tool would download this classifier instead of lists of trackers. There's a bit of poetic justice in using the tools of Big Data and machine learning, which are used by the tracking industry, to instead protect users against tracking.

5 Implications for Regulating Privacy

5.1 The Web Browser as a Privacy Regulator

As we saw in Sect. 2.3, the web browser mediates the user's interaction with web pages and trackers, and so browser vendors have considerable power over the state of online tracking. Vendors have been aware of this, but most have traditionally tended to remain "neutral". For example, browsers other than Safari have avoided blocking third-party cookies by default for this reason.

We view this stance as misguided. The web is so complex that the browser's defaults, user interface, and extension interface have an inevitable and tremendous impact on users' privacy outcomes. In the language of *Nudge*, the browser is a "choice architecture" and hence cannot be neutral [45]. In practice, attempts at neutrality have the effect of simply leaving in place the results of historical accidents. For example, the web standard explicitly leaves deciding how to handle third-party cookies to the browser, and most browsers made their original permissive decisions in a historical context where the privacy implications were not as clear as they are today.

[3]The Privacy Badger tool (https://www.eff.org/privacybadger) works somewhat like this, but it uses hard-coded heuristics instead of machine learning [44].

More recently, this attitude has been changing. Apple yielded to user demands to enable content blocking on Safari for iOS [46], Chrome is deliberating removing the filesystem API due to privacy-infringing use [47], and Microsoft enabled the Do Not Track signal by default in Internet Explorer 10 [48].[4] Firefox has been leading the charge, removing the Battery API due to abuse [49], enabling tracking protection in private browsing mode [50], and experimenting with other advanced privacy features [51].

We urge browser vendors to embrace their role as regulators of web privacy. There are important ongoing battles that pit consumer privacy against commercial interests, and browsers cannot and should not avoid taking a side. This will be especially tricky for browsers made by companies involved in online advertising. But browsers have the opportunity, through innovative technology, to steer the debate away from its current framing as a zero-sum game.

5.2 *Open Standards and Privacy*

The above discussion pertains to the baseline level of privacy on the web, i.e., the privacy outcome for the hypothetical average user. However, more technically skilled users may benefit from enabling optional browser features (including those squirreled away in "power user" interfaces) and installing and configuring browser extensions. In general, in an open platform, technologically savvy users are in a dramatically better position to protect their privacy and security, whereas in a closed platform, privacy and security outcomes are relatively uniform. Neither model is strictly better for privacy, but they do result in very different outcomes.

Inequality of privacy outcomes based on technical skill is worrying by itself, but also because such skill correlates with historically advantaged groups. It is a concern for researchers: new privacy technologies favor the tech savvy, so privacy research may *exacerbate* inequality in privacy outcomes unless combined with outreach efforts. This bias towards the tech savvy may also lead to a focus on technologies that are simply unsuitable for mainstream use, such as PGP. Inequality is especially a concern for the open-source model of privacy tool development, because in most cases there is no funding for usability testing or user education. There is a clear role for journalists, privacy activists, and civil society organizations to bridge the gap between developers and users of privacy tools. Consumer protection agencies could also play a role.

Policy makers should take note of the differences between open vs. closed platforms. As new platforms for data collection (such as the Internet of Things) take shape, it will be important to understand whether they lean open or closed. Open platforms present challenges for regulation since there isn't a natural point

[4]The Do Not Track standard itself lacks any teeth because of the failure of attempts at legislation or regulation to give it enforceable meaning.

of leverage, jurisdictional boundaries are harder to enforce, and the "long tail" of innovation makes enforcement difficult to scale. Given these challenges, one avenue for regulators is to complement technical measures, such as by clamping down on circumvention of cookie blocking [52]. We discuss two more approaches in the following two sections.

5.3 The Market for Lemons and First-Party Accountability

Many web publishers, as we have noted, have little awareness of the tracking on their own sites and the implications of it. The lack of oversight of third parties by publishers is a problem for privacy. This is most clearly seen in terms of the economic view that we used earlier. In a well-functioning market, many consumers will consider privacy (as one of several factors) in picking a product, service, website, etc. But in the case of online tracking, privacy failings are seen as the responsibility of third parties rather than publishers. Users don't interact directly with third parties, and therefore have no way to exercise their preferences in the marketplace.

We think this needs to change. When tracking is reported in the press, journalists should seek to make first parties accountable, and not just third parties. Similarly, privacy laws should make first parties primarily responsible for privacy violations. These will shift incentives so that first parties will start to have oversight of the tracking that happens on their domains. One of the very few examples that already embodies this principle is the US Children's Online Privacy Protection Act (COPPA), especially the Federal Trade Commission's COPPA rule [53] and recent enforcement actions against child-directed websites [54].

This transition won't be easy. Due to the financial struggles of publishers, regulators are loath to impose additional compliance burdens on them. But measurement tools can help. As part of WebTAP, we are building a publisher dashboard for a website operator to understand the tracking technologies in use on their own domain. Combined with the right technologies, the shift in incentives can in fact be a boon for publishers, who currently lack adequate information not just about tracking technologies but also about how tracking translates into revenue. In other words, our suggestion is to help shift the balance of power (and, with it, responsibility) from third parties to publishers.

5.4 Design for Measurement

Finally, our work suggests that policymakers have a lightweight way to intervene to improve privacy: by requiring service providers to support the ability of external researchers to measure and audit privacy. This could be as straightforward as the creation of APIs (application programming interfaces) for external measurement, to help automate studies that would otherwise require arduous manual effort. At the very least, policymakers should work to remove existing legal barriers to measurement research. Recently, the American Civil Liberties Union, together with academics, researchers, and journalists, have challenged the constitutionality of the Computer Fraud and Abuse Act on the grounds that it prevents uncovering racial discrimination online [55]. We welcome this development.

Transparency through external oversight is a valuable complement to—and sometimes more effective than—transparency through notice (i.e., privacy policies), for several reasons. Most importantly, external oversight doesn't require trusting the word of the company. Often, leaks of personally identifiable information (PII) or discriminatory effects of algorithms are introduced unintentionally into code, and measurement offers a way to discover these. Finally, measurement-based transparency can provide a precise and quantifiable view of privacy.

6 The Future of Transparency Research

Online tracking has proved an amenable target for large-scale measurement. Replicating the success of the WebTAP project in other domains will be much harder. For example, while there have been several interesting and important studies of privacy on smartphones and mobile devices, they don't reach the same scale and completeness, since app platforms are not as programmable as browsers [56–58] .

So far, WebTAP has looked primarily at data collection and data flows, and not nearly as much at uncovering bias and discrimination—and more generally, discovering how personal data is being used behind the scenes by algorithms. Here again it becomes harder to scale, because getting insights into personalization on a single website might require hundreds or thousands of observations. It also requires developing statistical techniques for establishing correlation or causation. Research teams at Columbia University and Carnegie Mellon University have recently made progress on this problem [59–62].

The most challenging scenarios for privacy measurement are also among the most important: those that involve the physical world. Think of the "Internet of Things" monitoring your activities in your home; tracking by analytics firms in shopping malls, based on WiFi and other emanations from your smartphone; apps that track your location and other activities; cross-device tracking and onboarding discussed in Sect. 4.3. The difficulty for researchers, of course, is that we can't quite automate the real world, at least not yet.

Researchers are adapting to these challenges in various ways. "Crowdsourcing" of data from different users' devices has resulted in important findings on privacy, price discrimination, and so on [8]. Sometimes it is possible to manipulate a fake user's location automatically [63, 64]. Engineers have created tools called "monkeys" that can simulate a user clicking through and exploring a smartphone app [65]. These were developed for finding bugs, but can also be used to automatically detect if the app phones home with personal data [66]. Computer science techniques such as static analysis and dynamic analysis allow analyzing or running an app in a simulated setting to understand its behavior [67]. Monitoring network communications generated by smartphones has proved powerful, especially combined with techniques to peek into encrypted traffic [68, 69]. Occasionally, researchers have rolled up their sleeves and conducted experiments manually in the absence of automated tools [70]. Finally, companies have sometimes been forthcoming in making provisions for researchers to study their systems.

That last point is important. Privacy research has too often been adversarial, and efforts by companies to be transparent and work with external researchers should be encouraged and rewarded. In conjunction, we need a science of designing systems to be transparent from the ground up. In the last few years, the Fairness, Accountability, and Transparency in Machine Learning ("FAT-ML") research community has made progress in developing algorithms that respect norms of transparency and non-discrimination.

While this progress is laudable, it appears that we have a long way to go in figuring out how to develop technology, especially machine learning, that respects our societal norms. A recent Princeton research paper (co-authored by Narayanan) looked for bias in machine learning—specifically, in a state-of-the-art technique called "word embeddings" that provide an algebraic representation of words that are easy for computers to manipulate [71]. The authors started from the "Implicit Association Test," a standard method in psychology to test human biases, and developed a version of it for word embeddings. They were able to replicate in the machine learning model every bias they tested that's been documented in humans, including racial and gender biases.

In other words, the underlying model of language used by the machine for a wide variety of tasks (such as language translation) is intrinsically biased. The authors argue that since these models are trained on text from the web written by humans, the machine inevitably absorbs the entire spectrum of human biases and prejudices, along with the rest of language, meaning and semantics. It is impossible to learn one without learning the other. This means that if we want to avoid enshrining our historical prejudices in our algorithms, we have to fundamentally re-examine the reigning paradigm of training machines on human knowledge and intelligence, and that will require a long-term research program.

7 Conclusion

The technology industry innovates at breakneck pace. But the more data-driven algorithms encroach into our lives, the more their complexity and inscrutability becomes problematic. A new area of empirical research seeks to make these systems more transparent, study their impact on society, and enable a modicum of external oversight. The Princeton Web Transparency and Accountability Project has focused on a small but important piece of this puzzle: third-party online tracking. By exploiting the open nature of web technologies, we have been able to "track the trackers" in an automated, large-scale, continual fashion, and to conduct a comprehensive study of tracking technologies used online.

An exciting but challenging future lies ahead for transparency research. Studying domains and systems that are less amenable to automated measurement will require various creative ideas. We hope these research findings will shape public policy, just as environmental policy is shaped by research examining the impact of human activities. Perhaps the greatest need is to develop a science of building data-driven algorithms in an ethical and transparent fashion from the ground up. Perhaps in the future algorithmic systems will even be built to explicitly support external measurement and oversight.

Acknowledgements Numerous graduate and undergraduate students and collaborators have contributed to the WebTAP project and to the findings reported here. In particular, Steven Englehardt is the primary student investigator and the lead developer of the OpenWPM measurement tool. We are grateful to Brian Kernighan, Vincent Toubiana, and the anonymous reviewer for useful feedback on a draft.

WebTAP is supported by NSF grant CNS 1526353, a grant from the Data Transparency Lab, and by Amazon AWS Cloud Credits for Research.

References

1. Crevier, D.: AI: The Tumultuous History of the Search for Artificial Intelligence. Basic Books, New York (1993)
2. Engle Jr., RL, Flehinger, B.J.: Why expert systems for medical diagnosis are not being generally used: a valedictory opinion. Bull. N. Y. Acad. Med. **63**(2), 193 (1987)
3. Vance, A.: This tech bubble is different. http://www.bloomberg.com/ (2011)
4. Angwin, J.: Machine bias: Risk assessments in criminal sentencing. ProPublica. https://www.propublica.org/ (2016)
5. Levin, S.: A beauty contest was judged by AI and the robots didn't like dark skin. https://www.theguardian.com/ (2016)
6. Solove, D.J.: Privacy and power: computer databases and metaphors for information privacy. Stanford Law Rev. **53**, 1393–1462 (2001)
7. Marthews, A., Tucker, C.: Government surveillance and internet search behavior. Available at SSRN 2412564 (2015)
8. Hannak, A., Soeller, G., Lazer, D., Mislove, A., Wilson, C.: Measuring price discrimination and steering on e-commerce web sites. In: Proceedings of the 2014 Conference on Internet Measurement Conference, pp. 305–318. ACM, New York (2014)
9. Calo, R.: Digital market manipulation. George Washington Law Rev. **82**, 995–1051 (2014)
10. Mayer, J.R., Mitchell, J.C.: Third-party web tracking: policy and technology. In: 2012 IEEE Symposium on Security and Privacy, pp. 413–427. IEEE, New York (2012)
11. Angwin, J.: The web's new gold mine: your secrets. ProPublica http://www.wsj.com/ (2010)
12. Lerner, A., Simpson, A.K., Kohno, T., Roesner, F.: Internet Jones and the raiders of the lost trackers: an archaeological study of web tracking from 1996 to 2016. In: 25th USENIX Security Symposium (USENIX Security 16) (2016)
13. Laperdrix, P., Rudametkin, W., Baudry, B.: Beauty and the beast: diverting modern web browsers to build unique browser fingerprints. In: 37th IEEE Symposium on Security and Privacy (S&P 2016) (2016)
14. Eckersley, P.: How unique is your web browser? In: International Symposium on Privacy Enhancing Technologies Symposium, pp. 1–18. Springer, Cambridge (2010)

15. Acar, G., Van Alsenoy, B., Piessens, F., Diaz, C., Preneel, B.: Facebook tracking through social plug-ins. Technical Report prepared for the Belgian Privacy Commission. https://securehomes.esat.kuleuven.be/~gacar/fb_tracking/fb_plugins.pdf (2015)
16. Starov, O., Gill, P., Nikiforakis, N.: Are you sure you want to contact us? quantifying the leakage of PII via website contact forms. In: Proceedings on Privacy Enhancing Technologies, vol. 2016(1), pp. 20–33 (2016)
17. Krishnamurthy, B., Naryshkin K, Wills C Privacy leakage vs. protection measures: the growing disconnect. In: Proceedings of the Web, vol. 2, pp. 1–10 (2011)
18. Su, J., Shukla, A., Goel, S., Narayanan, A.: De-anonymizing web browsing data with social networks, Manuscript (2017)
19. Barocas, S., Nissenbaum, H.: Big data's end run around procedural privacy protections. Commun. ACM **57**(11), 31–33 (2014)
20. Shilton, K., Greene, D.: Because privacy: defining and legitimating privacy in IOS development. In: Conference 2016 Proceedings (2016)
21. Storey, G., Reisman, D., Mayer, J., Narayanan, A.: The future of ad blocking: analytical framework and new techniques, Manuscript (2016)
22. Narayanan, A.: Can Facebook really make ads unblockable? https://freedom-to-tinker.com/ (2016)
23. Storey, G.: Facebook ad highlighter. https://chrome.google.com/webstore/detail/facebook-ad-highlighter/mcdgjlkefibpdnepeljmlfkbbbpkoamf?hl=en (2016)
24. Reisman, D.: A peek at A/B testing in the wild. https://freedom-to-tinker.com/ (2016)
25. Acar, G., Juarez, M., Nikiforakis, N., Diaz, C., Gürses, S., Piessens, F., Preneel, B.: Fpdetective: dusting the web for fingerprinters. In: Proceedings of the 2013 ACM SIGSAC Conference on Computer & Communications Security, pp. 1129–1140. ACM, New York (2013)
26. Englehardt, S., Narayanan, A.: Online tracking: a 1-million-site measurement and analysis. In: Proceedings of the 2016 ACM SIGSAC Conference on Computer & Communications Security (2016)
27. Selenium, H.Q.: Selenium browser automation FAQ. https://code.google.com/p/selenium/wiki/FrequentlyAskedQuestions (2016)
28. Acar, G., Eubank, C., Englehardt, S., Juarez, M., Narayanan, A., Diaz, C.: (2014) The web never forgets. In: Proceedings of the 2014 ACM SIGSAC Conference on Computer and Communications Security - CCS'14. doi:10.1145/2660267.2660347
29. Mowery, K., Shacham, H.: Pixel perfect: fingerprinting canvas in html5. In: Proceedings of W2SP (2012)
30. (Valve), V.V.: Fingerprintjs2 — modern & flexible browser fingerprinting library, a successor to the original fingerprintjs. https://github.com/Valve/fingerprintjs2 (2016)
31. Olejnik, Ł., Acar, G., Castelluccia, C., Diaz, C.: The leaking battery. In: International Workshop on Data Privacy Management, pp. 254–263. Springer, New York (2015)
32. Englehardt, S., Narayanan, A.: Online tracking: a 1-million-site measurement and analysis. In: Proceedings of the 2016 ACM SIGSAC Conference on Computer and Communications Security - CCS'16 (2016)
33. Doty, N.: Mitigating browser fingerprinting in web specifications. https://w3c.github.io/fingerprinting-guidance/ (2016)
34. Soltani, A., Peterson, A., Gellman, B.: NSA uses Google cookies to pinpoint targets for hacking. https://www.washingtonpost.com/ (2013)
35. Englehardt, S., Reisman, D., Eubank, C., Zimmerman, P., Mayer, J., Narayanan, A., Felten, E.W.: Cookies that give you away. In: Proceedings of the 24th International Conference on World Wide Web - WWW'15. doi:10.1145/2736277.2741679 (2015)
36. Angwin, J.: Google has quietly dropped ban on personally identifiable web tracking. ProPublica https://www.propublica.org (2016)
37. Reitman, R.: What actually changed in Google's privacy policy. Electronic Frontier Foundation https://www.eff.org (2012)
38. Simonite, T.: Facebook's like buttons will soon track your web browsing to target ads. MIT Technology Review https://www.technologyreview.com/ (2015)

39. Federal Trade Commission: Cross-device tracking. https://www.ftc.gov/news-events/events-calendar/2015/11/cross-device-tracking (2015)
40. Maggi, F., Mavroudis, V.: Talking behind your back attacks & countermeasures of ultrasonic cross-device tracking. https://www.blackhat.com/docs/eu-16/materials/eu-16-Mavroudis-Talking-Behind-Your-Back-Attacks-And-Countermeasures-Of-Ultrasonic-Cross-Device-Tracking.pdf, blackhat (2016)
41. Angwin, J.: Why online tracking is getting creepier. ProPublica https://www.propublica.org/ (2014)
42. Vallina-Rodriguez, N., Sundaresan, S., Kreibich, C., Paxson, V.: Header enrichment or ISP enrichment? Proceedings of the 2015 ACM SIGCOMM Workshop on Hot Topics in Middleboxes and Network Function Virtualization - HotMiddlebox'15 (2015). doi:10.1145/2785989.2786002
43. Disconnect: Disconnect blocks new tracking device that makes your computer draw a unique image. https://blog.disconnect.me/disconnect-blocks-new-tracking-device-that-makes-your-computer-draw-a-unique-image/ (2016)
44. Foundation, E.F.: Privacy badger. https://www.eff.org/privacybadger (2016)
45. Thaler, R.H., Sunstein, C.R.: Nudge: Improving Decisions About Health, Wealth, and Happiness. Yale University Press, New Haven (2008)
46. Fleishman, G.: Hands-on with content blocking safari extensions in iOS 9. Macworld http://www.macworld.com/ (2015)
47. Chromium-Blink: Owp storage team sync. https://groups.google.com/a/chromium.org/forum/#!topic/blink-dev/CT_eDVIdJv0 (2016)
48. Lynch, B.: Do not track in the windows 8 setup experience - microsoft on the issues. Microsoft on the Issues. https://blogs.microsoft.com/ (2012)
49. Hern, A.: Firefox disables loophole that allows sites to track users via battery status. The Guardian. https://www.theguardian.com/ (2016)
50. Mozilla: Tracking protection in private browsing. https://support.mozilla.org/en-US/kb/tracking-protection-pbm (2015)
51. Mozilla: Security/contextual identity project/containers. https://wiki.mozilla.org/Security/Contextual_Identity_Project/Containers (2016)
52. Federal Trade Commission: Google will pay $22.5 million to settle FTC charges it misrepresented privacy assurances to users of apple's safari internet browser. https://www.ftc.gov/news-events/press-releases/2012/08/google-will-pay-225-million-settle-ftc-charges-it-misrepresented (2012)
53. Federal Trade Commission: Children's online privacy protection rule ("coppa"). https://www.ftc.gov/enforcement/rules/rulemaking-regulatory-reform-proceedings/childrens-online-privacy-protection-rule (2016)
54. New York State Office of the Attorney General: A.G. schneiderman announces results of "operation child tracker," ending illegal online tracking of children at some of nation's most popular kids' websites. http://www.ag.ny.gov/press-release/ag-schneiderman-announces-results-operation-child-tracker-ending-illegal-online (2016)
55. American Civil Liberties Union: Sandvig v. Lynch. https://www.aclu.org/legal-document/sandvig-v-lynch-complaint-0 (2016)
56. Eubank, C., Melara, M., Perez-Botero, D., Narayanan, A.: Shining the floodlights on mobile web tracking – a privacy survey. http://www.w2spconf.com/2013/papers/s2p2.pdf (2013)
57. CMU CHIMPS Lab: Privacy grade: grading the privacy of smartphone apps. http://www.privacygrade.org (2015)
58. Vanrykel, E., Acar, G., Herrmann, M., Diaz, C.: Leaky birds: exploiting mobile application traffic for surveillance. In: Financial Cryptography and Data Security 2016 (2016)
59. Lécuyer, M., Ducoffe, G., Lan, F., Papancea, A., Petsios, T., Spahn, R., Chaintreau, A., Geambasu, R.: Xray: enhancing the web's transparency with differential correlation. In: 23rd USENIX Security Symposium (USENIX Security 14), pp. 49–64 (2014)
60. Lecuyer, M., Spahn, R., Spiliopolous, Y., Chaintreau, A., Geambasu, R., Hsu, D.: Sunlight: Fine-grained targeting detection at scale with statistical confidence. In: Proceedings of the 22nd ACM SIGSAC Conference on Computer and Communications Security, pp. 554–566. ACM, New York (2015)

61. Tschantz, M.C., Datta, A., Datta, A., Wing, J.M.: A methodology for information flow experiments. In: 2015 IEEE 28th Computer Security Foundations Symposium, pp. 554–568. IEEE, New York (2015)
62. Datta, A., Sen, S., Zick, Y.: Algorithmic transparency via quantitative input influence. In: Proceedings of 37th IEEE Symposium on Security and Privacy (2016)
63. Chen, L., Mislove, A., Wilson, C.: Peeking beneath the hood of Uber. In: Proceedings of the 2015 ACM Conference on Internet Measurement Conference, pp. 495–508. ACM, New York (2015)
64. Valentino-Devries, J., Singer-Vine, J., Soltani, A.: Websites vary prices, deals based on users' information. Wall Street J. (2012). https://www.wsj.com/articles/SB10001424127887323777204578189391813881534
65. Guide ASU: Ui/application exerciser monkey. https://developer.android.com/studio/test/monkey.html (2016)
66. Rastogi, V., Chen, Y., Enck, W.: AppsPlayground: automatic security analysis of smartphone applications. In: Proceedings of the Third ACM Conference on Data and Application Security and Privacy, pp. 209–220. ACM, New York (2013)
67. Enck, W., Gilbert, P., Han, S., Tendulkar, V., Chun, B.G., Cox, L.P., Jung, J., McDaniel, P., Sheth, A.N.: Taintdroid: an information-flow tracking system for realtime privacy monitoring on smartphones. ACM Trans. Comput. Syst. **32**(2), 5 (2014)
68. Ren, J., Rao, A., Lindorfer, M., Legout, A., Choffnes, D.: Recon: revealing and controlling privacy leaks in mobile network traffic (2015). arXiv preprint arXiv:150700255
69. Razaghpanah, A., Vallina-Rodriguez, N., Sundaresan, S., Kreibich, C., Gill, P., Allman, M., Paxson, V.: Haystack: in situ mobile traffic analysis in user space (2015). arXiv preprint arXiv:151001419
70. Sweeney, L.: Discrimination in online ad delivery. Queue **11**(3), 10 (2013)
71. Caliskan-Islam, A., Bryson, J., Narayanan, A.: Semantics derived automatically from language corpora necessarily contain human biases (2016). Arxiv https://arxiv.org/abs/1608.07187

Part II
Algorithmic Solutions

Algorithmic Transparency via Quantitative Input Influence

Anupam Datta, Shayak Sen, and Yair Zick

Abstract Algorithmic systems that employ machine learning are often opaque—it is difficult to explain why a certain decision was made. We present a formal foundation to improve the transparency of such decision-making systems. Specifically, we introduce a family of *Quantitative Input Influence (QII)* measures that capture the degree of input influence on system outputs. These measures provide a foundation for the design of *transparency reports* that accompany system decisions (e.g., explaining a specific credit decision) and for testing tools useful for internal and external oversight (e.g., to detect algorithmic discrimination). Distinctively, our *causal* QII measures carefully account for correlated inputs while measuring influence. They support a *general* class of transparency queries and can, in particular, explain decisions about individuals and groups. Finally, since single inputs may not always have high influence, the QII measures also quantify the *joint influence* of a set of inputs (e.g., age and income) on outcomes (e.g. loan decisions) and the *average marginal influence* of individual inputs within such a set (e.g., income) using principled aggregation measures, such as the Shapley value, previously applied to measure influence in voting.

Abbreviation

QII Quantitative Input Influence

A. Datta (✉) • S. Sen
Carnegie Mellon University, 5000 Forbes Avenue, Pittsburgh, PA, USA
e-mail: danupam@cmu.edu; shayaks@cmu.edu

Y. Zick
School of Computing, National University of Singapore, 21 Lower Kent Ridge Rd, Singapore, Singapore
e-mail: zick@comp.nus.edu.sg

T. Cerquitelli et al. (eds.), *Transparent Data Mining for Big and Small Data*, Studies in Big Data 32, DOI 10.1007/978-3-319-54024-5_4

Symbols

ι Influence
A Algorithm
Q_A Quantity of Interest

1 Introduction

Algorithmic decision-making systems that employ machine learning and related statistical methods are ubiquitous. They drive decisions in sectors as diverse as Web services, health-care, education, insurance, law enforcement and defense [2, 5, 6, 31, 34]. Yet their decision-making processes are often opaque. *Algorithmic transparency* is an emerging research area aimed at explaining decisions made by algorithmic systems.

The call for algorithmic transparency has grown in intensity as public and private sector organizations increasingly use large volumes of personal information and complex data analytics systems for decision-making [35]. Algorithmic transparency is important for several reasons. First, it is essential to enable identification of harms, such as discrimination, introduced by algorithmic decision-making (e.g., high interest credit cards targeted to protected groups) and to hold entities in the decision-making chain accountable for such practices. This form of accountability can incentivize entities to adopt appropriate corrective measures. Second, transparency can help detect errors in input data which resulted in an adverse decision (e.g., incorrect information in a user's profile because of which insurance or credit was denied). Such errors can then be corrected. Third, by explaining why an adverse decision was made, algorithmic transparency can provide guidance on how to reverse it (e.g., by identifying a specific factor in the credit profile that needs to be improved).

Our Goal While the importance of algorithmic transparency is recognized, work on computational foundations for this research area has been limited. In this direction, this chapter focuses on a concrete algorithmic transparency question:

> How can we measure the influence of inputs (or features) on decisions made by an algorithmic system about individuals or groups of individuals?

We wish to inform the design of transparency reports, which include answers to transparency queries of this form. To be concrete, let us consider a predictive policing system that forecasts future criminal activity based on historical data; individuals high on the list receive visits from the police. An individual who receives a visit from the police due to such an algorithmic decision can reasonably expect an explanation as to why. One possible explanation would be to provide a complete breakdown of the decision making process, or the source code of the program. However, this is unrealistic. Decision making algorithms are often very complex and protected as intellectual property.

Instead of releasing the entire source code of the decision making algorithm, our proposed transparency report provides answers to *personalized transparency queries* about the influence of various inputs (or features), such as race or recent criminal history, on the system's decision. These transparency reports can in turn be studied by oversight agencies or the general public in order to gain much needed insights into the use of sensitive user information (such as race or gender) by the decision making algorithm; as we later discuss, transparency reports can also point out potential systematic differences between groups of individuals (e.g. race-based discrimination). Our transparency reports can also be used to identify harms and errors in input data, and guide the design of better decision making algorithms.

Our Model We focus on a setting where a *transparency report* is generated with black-box access to the decision-making system[1] and knowledge of the input dataset on which it operates. This setting models the kind of access available to a private or public sector entity that proactively publishes transparency reports. It also models a useful level of access required for internal or external oversight of such systems to identify harms introduced by them. For the former use case, our approach provides a basis for design of transparency mechanisms; for the latter, it provides a formal basis for testing. Returning to our predictive policing example, the law enforcement agency that employs such a system could proactively publish transparency reports, and test the system for early detection of harms like race-based discrimination. An oversight agency could also use transparency reports for post hoc identification of harms.

Our Approach We formalize transparency reports by introducing a family of *Quantitative Input Influence (QII)* measures that capture the degree of influence of inputs on outputs of the system. Three desiderata drive the definitions of these measures.

First, we seek a formalization of a *general* class of transparency reports that allows us to answer many useful transparency queries related to input influence, including but not limited to the example forms described above about the system's decisions about individuals and groups.

Second, we seek input influence measures that appropriately account for *correlated inputs*—a common case for our target applications. For example, consider a system that assists in hiring decisions for a moving company. Gender and the ability to lift heavy weights are inputs to the system. They are positively correlated with one another and with the hiring decisions. Yet transparency into whether the system uses weight lifting ability or gender in making its decisions (and to what degree) has substantive implications for determining if it is engaging in discrimination (the business necessity defense could apply in the former case [16]). This observation makes us look beyond correlation coefficients and other associative measures.

[1]By "black-box access to the decision-making system" we mean a typical setting of software testing with complete control of inputs to the system and full observability of the outputs.

Third, we seek measures that appropriately quantify input influence in settings where any input by itself does not have significant influence on outcomes but a set of inputs does. In such cases, we seek measures of the *joint influence* of a set of inputs (e.g., age and income) on a system's decision (e.g., to serve a high-paying job ad). We also seek measures of the *marginal influence* of an input within such a set (e.g., age) on the decision. This notion allows us to provide finer-grained transparency about the relative importance of individual inputs within the set (e.g., age vs. income) in the system's decision.

We achieve the first desideratum by formalizing a notion of a *quantity of interest*. A transparency query measures the influence of an input on a quantity of interest, which represents a property of the system for a given input distribution. Our formalization supports a wide range of statistical properties including probabilities of various outcomes in the output distribution and probabilities of output distribution outcomes conditioned on input distribution events. Examples of quantities of interest include the conditional probability of an outcome for a particular individual or group, and the ratio of conditional probabilities for an outcome for two different groups (a metric used as evidence of disparate impact under discrimination law in the US [16]).

We achieve the second desideratum by formalizing *causal* QII measures. These measures (called *Unary QII*) model the difference in the quantity of interest when the system operates over two related input distributions—the real distribution and a hypothetical (or counterfactual) distribution that is constructed from the real distribution in a specific way to account for correlations among inputs. Specifically, if we are interested in measuring the influence of an input on a quantity of interest of the system behavior, we construct the hypothetical distribution by retaining the marginal distribution over all other inputs and sampling the input of interest from its prior distribution. This choice breaks the correlations between this input and all other inputs and thus lets us measure the influence of this input on the quantity of interest, independently of other correlated inputs. Revisiting our moving company hiring example, if the system makes decisions only using the weightlifting ability of applicants, the influence of gender will be zero on the ratio of conditional probabilities of being hired for males and females.

We achieve the third desideratum in two steps. First, we define a notion of joint influence of a set of inputs (called *Set QII*) via a natural generalization of the definition of the hypothetical distribution in the Unary QII definition. Second, we define a family of *Marginal QII* measures that model the difference on the quantity of interest as we consider sets with and without the specific input whose marginal influence we want to measure. Depending on the application, we may pick these sets in different ways, thus motivating several different measures. For example, we could fix a set of inputs and ask about the marginal influence of any given input in that set on the quantity of interest. Alternatively, we may be interested in the average marginal influence of an input when it belongs to one of several different sets that significantly affect the quantity of interest. We consider several marginal influence aggregation measures from cooperative game theory originally developed in the context of influence measurement in voting scenarios and discuss their applicability in our setting. We also build on that literature to present an efficient approximate algorithm for computing these measures.

Recognizing that different forms of transparency reports may be appropriate for different settings, we generalize our QII measures to be parametric in its key elements: the intervention used to construct the hypothetical input distribution; the quantity of interest; and the difference measure used to quantify the distance in the quantity of interest when the system operates over the real and hypothetical input distributions. This generalized definition provides a structure for exploring the design space of transparency reports.

We illustrate the utility of the QII framework by developing two machine learning applications on real datasets: an income classification application based on the benchmark `adult` dataset [27], and a predictive policing application based on the National Longitudinal Survey of Youth [32]. In particular, we analyze transparency reports for these two applications to demonstrate how QII can provide insights into individual and group classification outcomes.

2 Unary QII

Consider the situation discussed in the introduction, where an automated system assists in hiring decisions for a moving company. The input features used by this classification system are : *Age*, *Gender*, *Weight Lifting Ability*, *Marital Status* and *Education*. Suppose that, as before, weight lifting ability is strongly correlated with gender (with men having better overall lifting ability than women). One particular question that an analyst may want to ask is: "What is the influence of the input *Gender* on positive classification for women?". The analyst observes that 20% of women are approved according to his classifier. Then, she replaces every woman's field for gender with a random value, and notices that the number of women approved does not change. In other words, an *intervention* on the *Gender* variable does not cause a significant change in the classification outcome. Repeating this process with *Weight Lifting Ability* results in a 20% increase in women's hiring. Therefore, she concludes that for this classifier, *Weight Lifting Ability* has more influence on positive classification for women than *Gender*.

By breaking correlations between gender and weight lifting ability, we are able to establish a causal relationship between the outcome of the classifier and the inputs. We are able to identify that despite the strong correlation between a negative classification outcome for women, the feature *Gender* was not a cause of this outcome.

A mechanism for breaking correlations to identify causal effects is called an intervention in the literature on causality. In this chapter, we introduce a particular randomized intervention, and we describe in Sect. 3 how other interventions may be useful depending on the causal question asked of the system.

In the example above, instead of the entire population of women, the analyst may be interested in a particular individual, say Alice, and ask "What is the influence of the input *Gender* on Alice's rejection?". The analyst answers this question by applying the same intervention, but only on Alice's data, and observes that keeping every other input feature fixed and replacing *Gender* with a random value has no effect on the outcome.

In general, QII supports reasoning at such different scales by being parametric in a *quantity of interest*. A quantity of interest is a statistic of the system, which represents the subject of the causal question being asked. For example, a suitable quantity of interest corresponding to the question "What is the influence of the input *Gender* on positive classification for women?" is average rate of positive outcomes for women. Similarly, a suitable quantity of interest for "What is the influence of the input *Gender* on Alice's rejection?" is just Alice's classification outcome. QII also supports reasoning about more complex statistics such as ones which measure group disparity. For example, the analyst observes that the rate of positive classification is 20% higher in men than in women, and wishes to ask "What is the influence of *Income* on men getting more positive classification than women?". In this case, a natural quantity of interest is the difference in the classification rates of men and women.

We now present the formal definition of QII while omitting some technical details for which we refer the interested reader to the full technical paper [14].

We are given an algorithm $\mathscr{A}$. $\mathscr{A}$ operates on inputs (also referred to as *features* for ML systems), $N = \{1, \dots, n\}$. Every $i \in N$, can take on various *states*, given by X_i. We let $\mathscr{X} = \prod_{i \in N} \mathscr{X}_i$ be the set of possible feature state vectors, let $\mathscr{Z}$ be the set of possible outputs of $\mathscr{A}$. For a vector $\mathbf{x} \in \mathscr{X}$ and set of inputs $S \subseteq N$, $\mathbf{x}|_S$ denotes the vector of inputs in S. Inputs are assumed to be drawn from some probability distribution that represents the population. The random variable X represents a feature state vector drawn from this probability distribution.

For a random variable X representing the feature vector, a *randomized intervention* on feature i is denoted by the random variable $X_{-i}U_i$, where U_i is new random variable that is independently drawn from the prior distribution of feature i. For example, to randomly intervene on age, we replace the age for every individual with a random age from the population.

A *quantity of interest* $Q_{\mathscr{A}}(X)$ of some algorithm $\mathscr{A}$ is a statistic of the algorithm over the population X is drawn from. As discussed above, examples of quantities of interest are the average rate of positive classification for a group, or the classification outcome of an individual.

The Quantitative Input Influence of an input on a quantity of interest is just the difference in the quantity of interest with or without the intervention.

Definition 1 (QII) For a quantity of interest $Q_{\mathscr{A}}(\cdot)$, and an input i, the Quantitative Input Influence of i on $Q_{\mathscr{A}}(\cdot)$ is defined to be

$$\iota^{Q_{\mathscr{A}}}(i) = Q_{\mathscr{A}}(X) - Q_{\mathscr{A}}(X_{-i}U_i).$$

We now instantiate this definition with different quantities of interest to illustrate the above definition in three different scenarios.

QII for Individual Outcomes One intended use of QII is to provide personalized transparency reports to users of data analytics systems. For example, if a person is denied a job due to feedback from a machine learning algorithm, an explanation of which factors were most influential for that person's classification can provide valuable insight into the classification outcome.

For QII to quantify the use of an input for individual outcomes, we define the quantity of interest to be the classification outcome for a particular individual. Given a particular individual $\mathbf{x}$, we define $Q^{\mathbf{x}}_{\text{ind}}(\cdot)$ to be $\mathbb{E}(c(\cdot) = 1 \mid X = \mathbf{x})$. The influence measure is therefore:

$$\iota^{\mathbf{x}}_{\text{ind}}(i) = \mathbb{E}(c(X) = 1 \mid X = \mathbf{x}) - \mathbb{E}(c(X_{-i}U_i) = 1 \mid X = \mathbf{x}) \tag{1}$$

When the quantity of interest is not the probability of positive classification but the classification that $\mathbf{x}$ actually received, a slight modification of the above QII measure is more appropriate:

$$\begin{aligned} \iota^{\mathbf{x}}_{\text{ind-act}}(i) &= \mathbb{E}(c(X) = c(\mathbf{x}) \mid X = \mathbf{x}) - \mathbb{E}(c(X_{-i}U_i) = c(\mathbf{x}) \mid X = \mathbf{x}) \\ &= 1 - \mathbb{E}(c(X_{-i}U_i) = c(\mathbf{x}) \mid X = \mathbf{x}) = \mathbb{E}(c(X_{-i}U_i) \neq c(\mathbf{x}) \mid X = \mathbf{x}) \end{aligned} \tag{2}$$

The above probability can be interpreted as the probability that feature i is pivotal to the classification of $c(\mathbf{x})$. Computing the average of this quantity over X yields:

$$\textstyle\sum_{\mathbf{x}\in\mathcal{X}} \Pr(X = \mathbf{x})\mathbb{E}(i \text{ is pivotal for } c(X) \mid X = \mathbf{x}) = \mathbb{E}(i \text{ is pivotal for } c(X)). \tag{3}$$

We denote this average QII for individual outcomes as defined above, by $\iota_{\text{ind-avg}}(i)$, and use it as a measure for importance of an input towards classification outcomes.

QII for Group Outcomes For more general findings, the quantity of interest may be the classification outcome for a set of individuals. Given a group of individuals $\mathcal{Y} \subseteq \mathcal{X}$, we define $Q^{\mathcal{Y}}_{\text{grp}}(\cdot)$ to be $\mathbb{E}(c(\cdot) = 1 \mid X \in \mathcal{Y})$. The influence measure is therefore:

$$\iota^{\mathcal{Y}}_{\text{grp}}(i) = \mathbb{E}(c(X) = 1 \mid X \in \mathcal{Y}) - \mathbb{E}(c(X_{-i}U_i) = 1 \mid X \in \mathcal{Y}) \tag{4}$$

QII for Group Disparity Instead of simple classification outcomes, an analyst may be interested in more nuanced properties of data analytics systems such as the presence of disparate impact. Disparate impact compares the rates of positive classification within protected groups defined by gender or race. In employment, its absence is often codified as the '80% rule' which states that the rate of selection within a protected demographic should be at least 80% of the rate of selection within the unprotected demographic. The quantity of interest in such a scenario is the ratio in positive classification outcomes for a protected group $\mathcal{Y}$ from the rest of the population $\mathcal{X} \setminus \mathcal{Y}$.

$$\frac{\mathbb{E}(c(X) = 1 \mid X \in \mathcal{Y})}{\mathbb{E}(c(X) = 1 \mid X \notin \mathcal{Y})}$$

However, the ratio of classification rates is unstable at low values of positive classification. Therefore, for the computations in this paper we use the difference in classification rates as our measure of group disparity.

$$Q_{\text{disp}}^{\mathscr{Y}}(\cdot) = |\mathbb{E}(c(\cdot) = 1 \mid X \in \mathscr{Y}) - \mathbb{E}(c(\cdot) = 1 \mid X \notin \mathscr{Y})| \tag{5}$$

The QII measure of an input group disparity, as a result is:

$$\iota_{\text{disp}}^{\mathscr{Y}}(i) = Q_{\text{disp}}^{\mathscr{Y}}(X) - Q_{\text{disp}}^{\mathscr{Y}}(X_{-i}U_i). \tag{6}$$

More generally, group disparity can be viewed as an association between classification outcomes and membership in a group. QII on a measure of such association (e.g., group disparity) identifies the variable that causes the association in the classifier. *Proxy variables* are variables that are associated with protected attributes. For addressing concerns of discrimination such as *digital redlining*, it is important to identify which proxy variables actually introduce group disparity. It is straightforward to observe that features with high QII for group disparity are proxy variables that cause group disparity. Therefore, QII on group disparity is a useful diagnostic tool for determining discrimination. The use of QII in identifying proxy variables is explored experimentally in Sect. 5.1.

3 Transparency Schema

The forms of transparency reports discussed so far are specific instances of a general schema for algorithmic transparency via input influence that we describe now. Recall that the influence of an input is the difference in a quantity of interest before and after an intervention. This definition features the following three elements of our transparency schema.

- A *quantity of interest*, which captures the aspect of the system we wish to gain transparency into.
- An *intervention distribution*, which defines how the alternate distribution is constructed from the true distribution.
- A *difference measure*, which quantifies the difference between two quantities of interest.

The particular instantiation of elements in this schema is determined by the question the analyst wishes to answer about the system. Table 1 contains some examples of these instantiations and we describe the elements in detail below.

The first element, the *quantity of interest*, is determined by the subject of the question. In the previous section, the systems considered were deterministic, and the quantities of interest considered were represented by a number such as an expectation or a probability. However, for more complex systems, the analyst may be interested in richer quantities about the system. For instance, if the system is randomized, a particular quantity of interest is the distribution over outcomes for a given individual.

Table 1 Examples of transparency schema elements

Quantity of interest	Intervention	Difference
Individual outcome	Constant	Subtraction
Average outcome	Prior	Absolute difference
Distribution		Earthmover distance
Group disparity		Mutual information

The second element is the *causal intervention* used to compute influence. The particular randomized intervention considered in this paper, $X_{-i}U_i$, where U_i is drawn from the prior distribution of feature i, is suitable when no alternatives are suggested in the analyst's question. For example, the question "What is the influence of age on Bob's classification?" does not suggest any particular alternative age. On the other hand, when a particular alternative is suggested, a constant intervention is more suited. For example, for the question "Was age influential in Bob's rejection and Carl's acceptance?" Carl's age is the natural alternate to consider, and U_i is drawn from the constant distribution corresponding to Carl's age.

Finally, we use a *difference measure* to compare the quantity of interest before and after the intervention. Apart from subtraction, considered above, examples of other suitable difference measures are absolute difference, and distance measures over distributions. The absolute difference measure can be employed when the direction of change of outcomes before and after an intervention is unimportant. When the quantity of interest is the distribution over outcomes, subtraction does not make sense, and a suitable measure of difference should be employed, such as the earthmover distance or the mutual information between the original and the intervened distributions.

4 Set and Marginal QII

In many situations, intervention on a single input variable has no influence on the outcome of a system. Consider, for example, a two-feature setting where features are age (A) and income (I), and the classifier is $c(A, I) = (A = old) \wedge (I = high)$. In other words, the only datapoints that are labeled 1 are those of elderly persons with high income. Now, given a datapoint where $A = young, I = low$, an intervention on either age or income would result in the same classification. However, it would be misleading to say that neither age nor income have an influence over the outcome; changing both the states of income and age would result in a change in outcome.

Equating influence with the *individual* ability to affect the outcome is uninformative in real datasets as well: Fig. 1 is a histogram of feature influence on individual outcomes for a classifier learned from the `adult` dataset [27].[2]

[2]The `adult` dataset contains approximately 31k datapoints of users' personal attributes, and whether their income is more than $50k per annum; see Sect. 5 for more details.

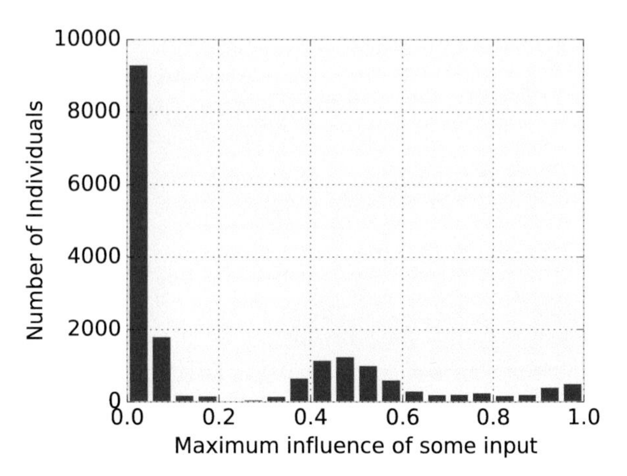

Fig. 1 A histogram of the highest specific causal influence for some feature across individuals in the adult dataset. Alone, most inputs have very low influence

For most individuals, all features have zero influence: changing the state of one feature alone is not likely to change the outcome of a classifier. Of the 19,537 datapoints we evaluate, more than half have $\iota^{\mathbf{x}}(i) = 0$ for all $i \in N$; indeed, changes to outcome are more likely to occur if we intervene on *sets of features*. In order to better understand the influence of a feature $i \in N$, we should measure its effect when coupled with interventions on other features. We define the influence of a set of inputs as a straightforward extension of individual input influence. Essentially, we wish the influence of a set of inputs $S \subseteq N$ to be the same as when the set of inputs is considered to be a single input; when intervening on S, we draw the states of $i \in S$ based on the joint distribution of the states of features in S.

We define the random variable $X_{-S}U_S$ representing an intervention on a set of features instead of a single feature. $X_{-S}U_S$ has the states of features in $N \setminus S$ fixed to their original values in $\mathbf{x}$, but features in S take on new values according the joint marginal distribution over the features in S. This is identical to the intervention considered in unary QII, when the features in S are viewed as a single composite feature. Using this generalization of an intervention to sets of features, we can define set QII that measures the influence of a set of features.

Definition 2 (Set QII) For a quantity of interest Q, and an input i, the Quantitative Input Influence of set $S \subseteq N$ on Q is defined to be

$$\iota^Q(S) = Q(X) - Q(X_{-S}U_S).$$

Considering the influence of a set of inputs raises a number of interesting questions due to the interaction between inputs. First among these is how does one measure the *individual effect* of a feature, given the measured effects of interventions on sets of features. One natural way of doing so is by measuring the *marginal effect* of a feature on a set.

Definition 3 (Marginal QII) For a quantity of interest Q, and an input i, the Quantitative Input Influence of input i over a set $S \subseteq N$ on Q is defined to be

$$\iota^Q(i, S) = Q(X_{-S}U_S) - Q(X_{-S\cup\{i\}}U_{S\cup\{i\}}).$$

Notice that marginal QII can also be viewed as a difference in set QIIs: $\iota^Q(S \cup \{i\}) - \iota^Q(S)$. Informally, the difference between $\iota^Q(S \cup \{i\})$ and $\iota^Q(S)$ measures the "added value" obtained by intervening on $S \cup \{i\}$, versus intervening on S alone.

The marginal contribution of i may vary significantly based on S. Thus, we are interested in the *aggregate marginal contribution* of i to S, where S is sampled from some natural distribution over subsets of $N \setminus \{i\}$. In what follows, we describe a few measures for aggregating the marginal contribution of a feature i to sets, based on different methods for sampling sets. The primary method of aggregating the marginal contribution is the Shapley value [39]. The less theoretically inclined reader can choose to proceed to Sect. 5 without a loss in continuity.

4.1 Cooperative Games and Causality

In this section, we discuss how measures from the theory of cooperative games define measures for aggregating marginal influence. In particular, we observe that the Shapley value [39] is uniquely characterized by axioms that are natural in our setting. Definition 2 measures the influence that an intervention on a set of features $S \subseteq N$ has on the outcome. One can naturally think of Set QII as a function $v : 2^N \to \mathbb{R}$, where $v(S)$ is the influence of S on the outcome. With this intuition in mind, one can naturally study influence measures using *cooperative game theory*, and in particular, using the Shapley value. The Shapley value can be thought of as an *influence aggregation method*, which, given an influence measure $v : 2^N \to \mathbb{R}$, outputs a vector $\phi \in \mathbb{R}^n$, whose i-th coordinate corresponds in some natural way to the aggregate influence, or aggregate causal effect, of feature i.

The original motivation for game-theoretic measures is *revenue division* [29, Chapter 18]: the function v describes the amount of money that each subset of players $S \subseteq N$ can generate; assuming that the set N generates a total revenue of $v(N)$, how should $v(N)$ be divided amongst the players? A special case of revenue division that has received significant attention is the measurement of voting power [40]. In voting systems with multiple agents (think of political parties in a parliament), whose weights differ, voting power often does not directly correspond to agent weights. For example, the US presidential election can roughly be modeled as a cooperative game where each state is an agent. The weight of a state is the number of electors in that state (i.e., the number of votes it brings to the presidential candidate who wins that state). Although states like California and Texas have higher weight, swing states like Pennsylvania and Ohio tend to have higher power in determining the outcome of elections.

A voting system can be modeled as a cooperative game: players are voters, and the value of a coalition $S \subseteq N$ is 1 if S can make a decision (e.g. pass a bill, form a government, or perform a task), and is 0 otherwise. Note the similarity to classification, with players being replaced by features (in classification, functions of the form $v : 2^N \to \{0, 1\}$ are referred to as boolean functions [33]). The game-theoretic measures of revenue division are a measure of *voting power*: how much influence does player i have in the decision making process? Thus the notions of voting power and revenue division fit naturally with our goals when defining aggregate QII influence measures: in both settings, one is interested in measuring the aggregate effect that a single element has, given the actions of subsets.

Several canonical influence measures rely on the fundamental notion of *marginal contribution*. Given a player i and a set $S \subseteq N \setminus \{i\}$, the marginal contribution of i to S is denoted $m_i(S, v) = v(S \cup \{i\}) - v(S)$ (we simply write $m_i(S)$ when v is clear from the context). Marginal QII, as defined above, can be viewed as an instance of a measure of marginal contribution. Given a permutation $\pi \in \Pi(N)$ of the elements in N, we define $P_i(\sigma) = \{j \in N \mid \sigma(j) < \sigma(i)\}$; this is the set of i's *predecessors* in σ. We can now similarly define the marginal contribution of i to a permutation $\sigma \in \Pi(N)$ as $m_i(\sigma) = m_i(P_i(\sigma))$. Intuitively, one can think of the players sequentially entering a room, according to some ordering σ; the value $m_i(\sigma)$ is the marginal contribution that i has to whoever is in the room when she enters it.

Generally speaking, game theoretic influence measures specify some reasonable way of aggregating the marginal contributions of i to sets $S \subseteq N$. In our setting, we argue for the use of the Shapley value. Introduced by the late Lloyd Shapley, the Shapley value is one of the most canonical methods of dividing revenue in cooperative games. It is defined as follows:

$$\varphi_i(N, v) = \mathbb{E}_\sigma[m_i(\sigma)] = \frac{1}{n!} \sum_{\sigma \in \Pi(N)} m_i(\sigma)$$

Intuitively, the Shapley value describes the following process: players are sequentially selected according to some randomly chosen order σ; each player receives a payment of $m_i(\sigma)$. The Shapley value is the expected payment to the players under this regime.

4.2 Axiomatic Treatment of the Shapley Value

In this work, the Shapley value is our function of choice for aggregating marginal feature influence. To justify our choice, we provide a brief exposition of axiomatic game-theoretic value theory. We present a set of axioms that uniquely define the Shapley value, and discuss why they are desirable in the QII setting.

The Shapley value satisfies the following properties:

Definition 4 (Symmetry) We say that $i, j \in N$ are *symmetric* if $v(S \cup \{i\}) = v(S \cup \{j\})$ for all $S \subseteq N \setminus \{i, j\}$. A value ϕ satisfies *symmetry* if $\phi_i = \phi_j$ whenever i and j are symmetric.

Definition 5 (Dummy) We say that a player $i \in N$ is a *dummy* if $v(S \cup \{i\}) = v(S)$ for all $S \subseteq N$. A value ϕ satisfies the *dummy* property if $\phi_i = 0$ whenever i is a dummy.

Definition 6 (Monotonicity) Given two games $\langle N, v_1 \rangle, \langle N, v_2 \rangle$, a value ϕ satisfies *strong monotonicity* if $m_i(S, v_1) \geq m_i(S, v_2)$ for all S implies that $\phi_i(N, v_1) \geq \phi_i(N, v_2)$, where a strict inequality for some set $S \subseteq N$ implies a strict inequality for the values as well.

All of these axioms take on a natural interpretation in the QII setting. Indeed, if two features have the same probabilistic effect, no matter what other interventions are already in place, they should have the same influence. In our context, the dummy axiom says that a feature that never offers information with respect to an outcome should have no influence. Monotonicity makes intuitive sense in the QII setting: if a feature has consistently higher influence on the outcome in one setting than another, its measure of influence should increase. For example, if a user receives two transparency reports (say, for two separate loan applications), and in one report gender had a consistently higher effect on the outcome than in the other, then the transparency report should reflect this.

[47] offers an characterization of the Shapley value, based on these *monotonicity* assumption.

Theorem 1 ([47]) *The Shapley value is the only function that satisfies symmetry, dummy and monotonicity.*

To conclude, the Shapley value is a *unique* way of measuring aggregate influence in the QII setting, while satisfying a set of very natural axioms.

5 Experimental Evaluation

We illustrate the utility of the QII framework by developing two simple machine learning applications on real datasets. In Sect. 5.1, we illustrate the distinction between different quantities of interest on which Unary QII can be computed. We also illustrate the effect of discrimination on the QII measure. In Sect. 5.2, we analyze transparency reports of three individuals to demonstrate how Marginal QII can provide insights into individuals' classification outcomes.

We use the following datasets in our experiments:

- `adult`[27]: This standard machine learning benchmark dataset is a subset of US census data classifying individual income; it contains factors such as age, race, gender, marital status and other socio-economic parameters. We use this dataset to train a classifier that predicts the income of individuals from other parameters. Such a classifier could potentially be used to assist in credit decisions.

- `arrests`[32]: The National Longitudinal Surveys are a set of surveys conducted by the Bureau of Labor Statistics of the United States. In particular, we use the 1997 youth survey, covering young persons born in the years 1980–1984. The survey covers various aspects of an individual's life such as medical history, criminal records and economic parameters. From this dataset, we extract the following features: age, gender, race, region, history of drug use, history of smoking, and history of arrests. We use this data to train a classifier that predicts arrests history to simulate predictive policing, where socio-economic factors are used to decide whether individuals should receive a visit from the police. A similar application is described in [21].

The two applications described above are hypothetical examples of decision making aided by algorithms that use potentially sensitive socio-economic data, and not real systems that are currently in use. We use these classifiers to illustrate the subtle causal questions that our QII measures can answer.

We use the following standard machine learning classifiers in our dataset: Logistic Regression, SVM with a radial basis function kernel, Decision Tree, and Gradient Boosted Decision Trees (see [7] for an excellent primer to these classifiers); with the exception of Logistic Regression, a linear classifier, the other three are nonlinear and can potentially learn very complex models.

5.1 Unary QII Measures

In Fig. 2, we illustrate the use of different Unary QII measures. Figure 2a, b show the Average QII measure (Eq. (3)) computed for features of a decision forest classifier. For the income classifier trained on the `adult` dataset, the feature with highest influence is *Marital Status*, followed by *Occupation*, *Relationship* and *Capital Gain*. Sensitive features such as *Gender* and *Race* have relatively lower influence. For the predictive policing classifier trained on the `arrests` dataset, the most influential input is *Drug History*, followed by *Gender*, and *Smoking History*. We observe that influence on outcomes may be different from influence on group disparity.

QII on Group Disparity Figure 2c, d show influences of features on group disparity (Eq. (6)) for two different settings. Figure 2c shows feature influence on group disparity by Gender in the `adult` dataset; Fig. 2d shows the influence on group disparity by Race in the `arrests` dataset. For the income classifier trained on the `adult` dataset, we observe that most inputs have negative influence on group disparity; randomly intervening on most inputs would lead to a reduction in group disparity. In other words, a classifier that did not use these inputs would be fairer. Interestingly, in this classifier, marital status—rather than gender—has the highest influence on group disparity by gender.

For the `arrests` dataset, most inputs have the effect of increasing group disparity if randomly intervened on. In particular, *Drug history* has the highest positive influence on disparity in `arrests`. Although Drug history is correlated with race, using it reduces disparate impact by race, i.e. makes fairer decisions.

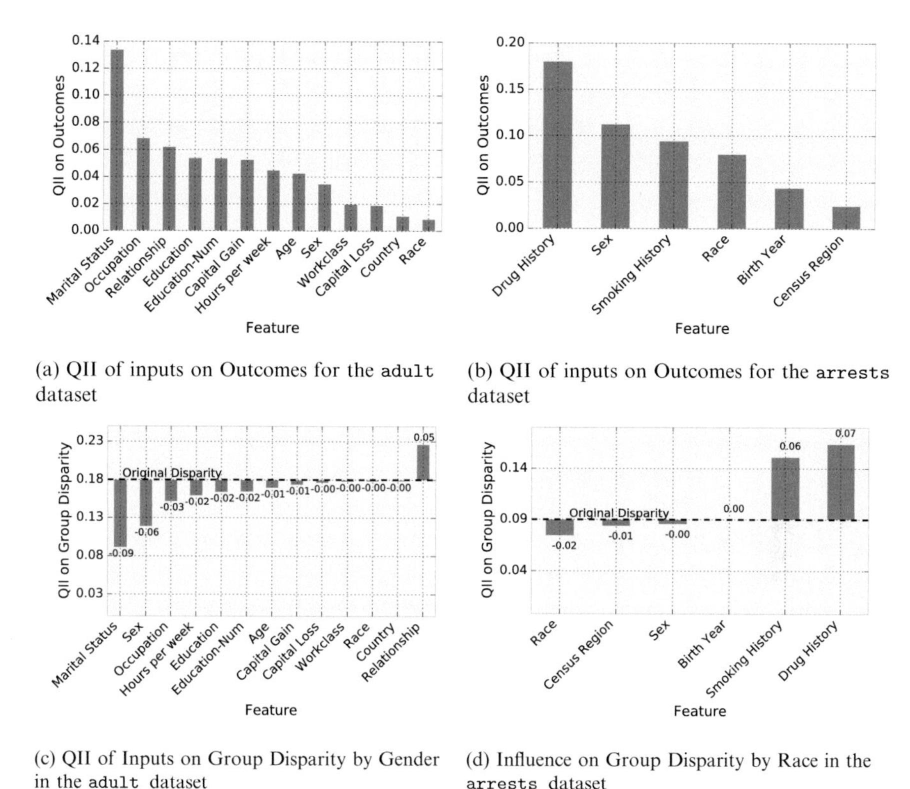

(a) QII of inputs on Outcomes for the adult dataset

(b) QII of inputs on Outcomes for the arrests dataset

(c) QII of Inputs on Group Disparity by Gender in the adult dataset

(d) Influence on Group Disparity by Race in the arrests dataset

Fig. 2 QII measures for the adult and arrests datasets. (**a**) QII of inputs on outcomes for the adult dataset, (**b**) QII of inputs on Outcomes for the arrests dataset, (**c**) QII of Inputs on group disparity by gender in the adult dataset, (**d**) Influence on group disparity by race in the arrests dataset

In both examples, features correlated with the sensitive attribute are the most influential for group disparity according to the sensitive attribute rather than the sensitive attribute itself. It is in this sense that QII measures can identify proxy variables that cause associations between outcomes and sensitive attributes.

QII with Artificial Discrimination We simulate discrimination using an artificial experiment. We first randomly assign "ZIP codes" to individuals in our dataset (these are simply arbitrary numerical identifiers). Then, to simulate systematic bias, we make an f fraction of the ZIP codes discriminatory in the following sense: all individuals in the protected set are automatically assigned a negative classification outcome. We then study the change in the influence of features as we increase f. Figure 3a, shows that the influence of *Gender* increases almost linearly with f. Recall that *Marital Status* was the most influential feature for this classifier without any added discrimination. As f increases, the importance of *Marital Status* decreases as expected, since the number of individuals for whom *Marital Status* is pivotal decreases.

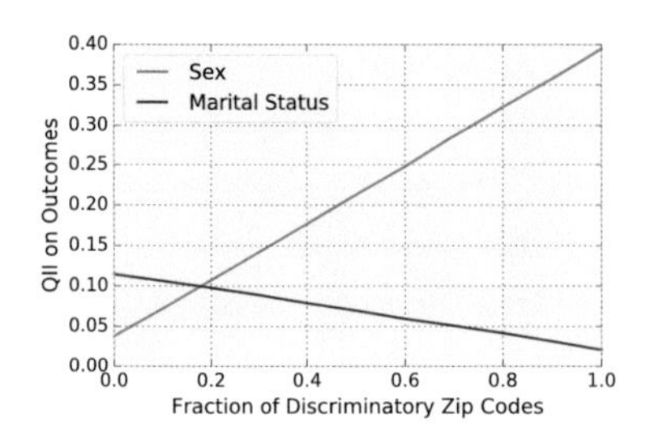

(a) Change in QII of inputs as discrimination by Zip Code increases in the `adult` dataset

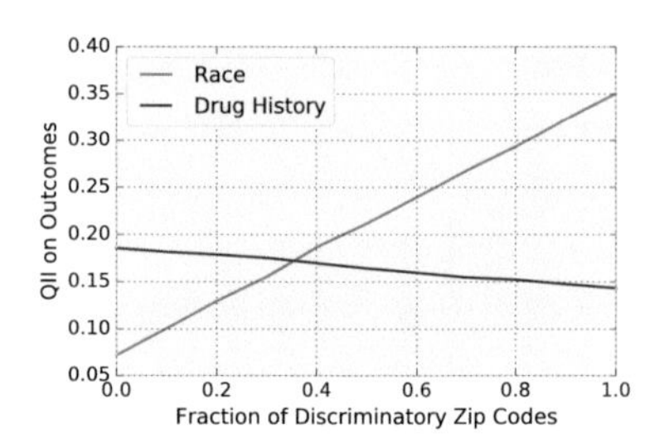

(b) Change in QII of inputs as discrimination by Zip Code increases in the `arrests` dataset

Fig. 3 The effect of discrimination on QII. (**a**) Change in QII of inputs as discrimination by Zip code increases in the `adult` dataset, (**b**) Change in QII of inputs as discrimination by Zip code increases in the `arrests` dataset

5.2 *Personalized Transparency Reports*

To illustrate the utility of personalized transparency reports, we study the classification of individuals who received potentially unexpected outcomes. For the personalized transparency reports, we use classification outcomes obtained from decision forests, though one obtains a similar outcome using other classifiers.

The influence measure that we employ is the Shapley value. In more detail, the coalitional function we use is $v(S) = \iota^{Q_{\mathscr{A}}}(S)$, with $Q_{\mathscr{A}}$ being $\mathbb{E}[c(\cdot) = 1 \mid X = \mathbf{x}]$; that is, the marginal contribution of $i \in N$ to S is given by $m_i(S) = \mathbb{E}[c(X_{-S}) = 1 \mid X = \mathbf{x}] - \mathbb{E}[c(X_{-S\cup\{i\}}) = 1 \mid X = \mathbf{x}]$.

We emphasize that some features may have a negative Shapley value; this should be interpreted as follows: a feature with a high positive Shapley value often increases the certainty that the classification outcome is 1, whereas a feature whose Shapley value is negative is one that increases the certainty that the classification outcome would be zero.

Mr. X: The first example is of an individual from the `adult` dataset, to whom we refer as Mr. X (Fig. 4). The learnt classifier classifies his income as low. This result may be surprising to him: he reports high capital gains (\$14k), and only 2.1% of people with capital gains higher than \$10k are reported as low income. In fact, he might be led to believe that his classification may be a result of his ethnicity or country of origin. Examining his transparency report in Fig. 4, however, we find that the most influential features that led to his negative classification were Marital Status, Relationship and Education.

Mr. Y: The second example, to whom we refer as Mr. Y (Fig. 5), has even higher capital gains than Mr. X. Mr. Y is a 27 year old, with only Preschool education, and is engaged in fishing. Examination of the transparency report reveals that the most influential factor for negative classification for Mr. Y is his Occupation. Interestingly, his low level of education is not considered very important by this classifier.

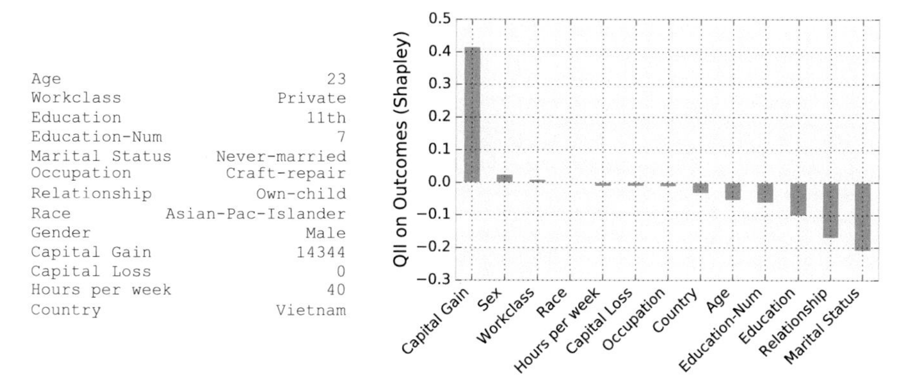

Age	23
Workclass	Private
Education	11th
Education-Num	7
Marital Status	Never-married
Occupation	Craft-repair
Relationship	Own-child
Race	Asian-Pac-Islander
Gender	Male
Capital Gain	14344
Capital Loss	0
Hours per week	40
Country	Vietnam

Fig. 4 Mr. X's profile and transparency report for negative classification

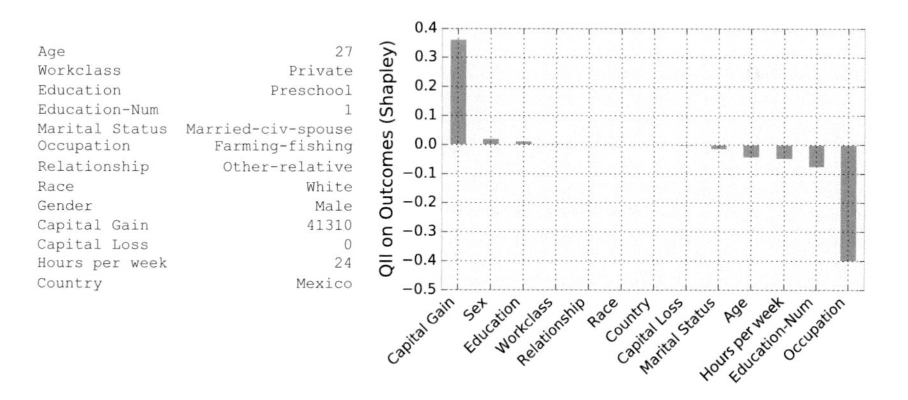

Age	27
Workclass	Private
Education	Preschool
Education-Num	1
Marital Status	Married-civ-spouse
Occupation	Farming-fishing
Relationship	Other-relative
Race	White
Gender	Male
Capital Gain	41310
Capital Loss	0
Hours per week	24
Country	Mexico

Fig. 5 Mr. Y's profile and transparency report for negative classification

Mr. Z: The third example, who we refer to as Mr. Z (Fig. 6) is from the `arrests` dataset. History of drug use and smoking are both strong indicators of arrests. However, Mr. X received positive classification by this classifier even without any history of drug use or smoking. Upon examining his classifier, it appears that race, age and gender were most influential in determining his outcome. In other words, the classifier that we train for this dataset (a decision forest) has picked up on the correlations between race (Black), and age (born in 1984) to infer that this individual is likely to engage in criminal activity. Indeed, our interventional approach indicates that this is not a mere correlation effect: race is actively being used by this classifier to determine outcomes. Of course, in this instance, we have explicitly offered the race parameter to our classifier as a viable feature. However, our influence measure is able to pick up on this fact, and alert us of the problematic behavior of the underlying classifier. More generally, this example illustrates a concern with the black box use of machine learning which can lead to unfavorable outcomes for individuals.

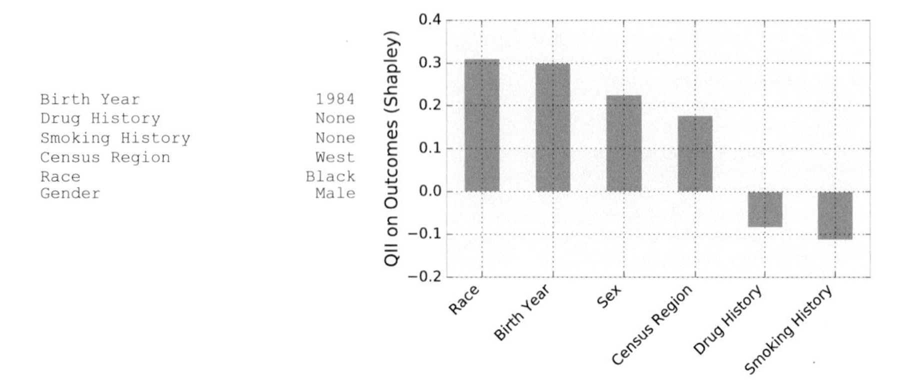

Fig. 6 Mr. Z's profile and transparency report for positive classification

6 QII for Fairness

Due to the widespread and black box use of machine learning in aiding decision making, there is a legitimate concern of algorithms introducing and perpetuating social harms such as racial discrimination [4, 35]. As a result, the algorithmic foundations of fairness in personal information processing systems have received significant attention recently [10, 12, 15, 22, 48]. While many of the algorithmic approaches [10, 17, 22, 48] have focused on group parity as a metric for achieving fairness in classification, Dwork et al. [15] argue that group parity is insufficient as a basis for fairness, and propose an approach which prescribes that similar individuals should receive similar classification outcomes. However, this approach requires a similarity metric for individuals which is often subjective and difficult to construct.

QII does not suggest any normative definition of fairness. Instead, we view QII as a diagnostic tool to aid fine-grained fairness determinations. In fact, QII can be used in the spirit of the similarity based definition of [15] by comparing the personalized privacy reports of individuals who are *perceived* to be similar but received different classification outcomes, and identifying the inputs which were used by the classifier to provide different outcomes.

When group parity is used as a criteria for fairness, QII can identify the features that lead to group disparity, thereby identifying features being used by a classifier as a proxy for sensitive attributes. The determination of whether using certain proxies for sensitive attributes is discriminatory is often a task-specific normative judgment. For example, using standardized test scores (e.g., SAT scores) for college admissions decisions is by and large accepted. However, SAT scores may act as a proxy for several protected attributes, leading to concerns of unfairness [43, 45]. Our goal is not to provide such normative judgments. Rather we seek to provide fine-grained transparency into input usage (e.g., what's the extent to which SAT scores influence decisions), which is useful to make determinations of discrimination from a specific normative position.

7 Related Work

QII enables algorithmic transparency via a family of game-theoretic causal influence measures. In this section we compare with related work along a number of different dimensions. First, we compare with related approaches to measuring causal influence, associations, and briefly discuss an emerging direction that measures proxy influence via a combination of associative and causal techniques. Next, we discuss orthogonal approaches to algorithmic transparency: interpretable machine learning and experimentation on web services. Finally, we discuss other fields in which similar game-theoretic measures have been applied.

Quantitative Causal Measures Causal models and probabilistic interventions have been used in a few other settings. While the form of the interventions in some of these settings may be very similar, our generalization to account for different quantities of interests enables us to reason about a large class of transparency queries for data analytics systems ranging from classification outcomes of individuals to disparity among groups. Further, the notion of marginal contribution which we use to compute responsibility does not appear in this line of prior work.

Janzing et al. [20] use interventions to assess the causal importance of relations between variables in causal graphs; in order to assess the causal effect of a relation between two variables, $X \rightarrow Y$ (assuming that both take on specific values $X = x$ and $Y = y$), a new causal model is constructed, where the value of X is replaced with a prior over the possible values of X. The influence of the causal relation is defined as the distance between the joint distributions of all the variables in the two causal models with and without the value of X replaced. The approach of intervening with a random value from the prior is similar to our approach of constructing $X_{-S}U_S$.

Independently, there has been considerable work in the machine learning community to define importance metrics for variables, mainly for the purpose of feature selection (see [19] for a comprehensive overview). One important metric is called Permutation Importance [9], which measures the importance of a feature towards classification by randomly permuting the values of the feature and then computing the difference of classification accuracies before and after the permutation. Replacing a feature with a random permutation can be viewed as sampling the feature independently from the prior.

Measures of Association One can think of our results as a causal alternative to *quantitative information flow*. Quantitative information flow is a broad class of metrics that quantify the information leaked by a process by comparing the *information* contained before and after observing the outcome of the process. Quantitative Information Flow traces its information-theoretic roots to the work of Shannon [38] and Rényi [36]. Recent works have proposed measures for quantifying the security of information by measuring the amount of information leaked from inputs to outputs by certain variables; we point the reader to [41] for an overview, and to [11] for an exposition on information theory. Quantitative Information Flow is concerned with information leaks and therefore needs to account for correlations between inputs that may lead to leakage. The dual problem of transparency, on the other hand, requires us to destroy correlations while analyzing the outcomes of a system to identify the causal paths for information leakage. Measures of association have also been widely used for detecting discrimination (see Sect. 6).

Proxy Influence An emerging direction in this space is the identification of proxy or indirect use of sensitive features as opposed to direct causal use, as captured by QII. Adler et al. [1] quantify the indirect influence of an attribute by obscuring the attribute (along with associations) from a dataset and comparing the prediction accuracy of a model before and after obscuring. In a different approach, Datta et al. (Use privacy in data-driven systems: theory and experiments with machine learnt programs, Unpublished Manuscript) identify intermediate computations in a program that are associated with an attribute and use QII to measure the causal influence of the intermediate computations on the outcome.

Interpretable Machine Learning An orthogonal approach to adding interpretability to machine learning is to constrain the choice of models to those that are interpretable by design. This can either proceed through regularization techniques such as Lasso [44] that attempt to pick a small subset of the most important features, or by using models that structurally match human reasoning such as Bayesian Rule Lists [26], Supersparse Linear Integer Models [46], or Probabilistic Scaling [37]. Since the choice of models in this approach is restricted, a loss in predictive accuracy is a concern, and therefore, the central focus in this line of work is the minimization of the loss in accuracy while maintaining interpretability. On the other hand, our approach to interpretability is forensic. We add interpretability to machine learning models after they have been learnt. As a result, our approach does not constrain the choice of models that can be used.

Experimentation on Web Services Another emerging body of work focuses on systematic experimentation to enhance transparency into Web services such as targeted advertising [3, 12, 18, 24, 25]. The setting in this line of work is different since they have restricted access to the analytics systems through publicly available interfaces. As a result they only have partial control of inputs, partial observability of outputs, and little or no knowledge of input distributions. The intended use of these experiments is to enable external oversight into Web services without any cooperation. Our framework is more appropriate for a transparency mechanism where an entity proactively publishes transparency reports for individuals and groups. Our framework is also appropriate for use as an internal or external oversight tool with access to mechanisms with control and knowledge of input distributions, thereby forming a basis for testing.

Game-Theoretic Influence Measures Recent years have seen game-theoretic influence measures used in various settings. Datta et al. [13] also define a measure for quantifying feature influence in classification tasks. Their measure does not account for the prior on the data, nor does it use interventions that break correlations between sets of features. In the terminology of this paper, the quantity of interest used by Datta et al. [13] is the ability of changing the outcome by changing the state of a feature. Strumbelj and Kononenko [42] also use the Shapley value to compute influences for individual classification. This work greatly extends and generalizes the concepts presented in [13] and [42], by both accounting for interventions on sets, and by generalizing the notion of influence to include a wide range of system behaviors, such as group disparity, group outcomes and individual outcomes. Essentially, the instantiations of our transparency schema define a wide range of transparency reports.

Game theoretic measures have been used by various research disciplines to measure influence. Indeed, such measures are relevant whenever one is interested in measuring the marginal contribution of variables, and when sets of variables are able to cause some measurable effect. Prominent domains where game theoretic measures have recently been used are terrorist networks [28, 30], protein interactions [8], and neurophysical models [23]. The novelty in our use of the game theoretic power indices lies in the conception of a cooperative game via a valuation function $\iota(S)$, defined by a randomized intervention on input S. Such an intervention breaks correlations and allows us to compute marginal causal influences on a wide range of system behaviors.

8 Conclusion & Future Work

In this chapter, we present QII, a general family of metrics for quantifying the influence of inputs in systems that process personal information. In particular, QII lends insights into the behavior of opaque machine learning algorithms by allowing us to answer a wide class of transparency queries ranging from influence on individual causal outcomes to influence on disparate impact. To achieve this, QII breaks correlations between inputs to allow causal reasoning, and computes the marginal influence of inputs in situations where inputs cannot affect outcomes alone. Also, we demonstrate that QII can be efficiently approximated, and can be made differentially private with negligible noise addition in many cases.

We do not consider situations where inputs do not have well understood semantics. Such situations arise often in settings such as image or speech recognition, and automated video surveillance. With the proliferation of immense processing power, complex machine learning models such as deep neural networks have become ubiquitous in these domains. Similarly, we do not consider the problem of measuring the influence of inputs that are not explicitly provided but inferred by the model. Defining transparency and developing analysis techniques in such settings is important future work.

References

1. Adler, P., Falk, C., Friedler, S., Rybeck, G., Schedegger, C., Smith, B., Venkatasubramanian, S.: Auditing black-box models for indirect influence. In: Proceedings of the 2016 IEEE International Conference on Data Mining (ICDM), ICDM'16, pp. 339–348. IEEE Computer Society, Washington (2016)
2. Alloway, T.: Big data: Credit where credit's due (2015). http://www.ft.com/cms/s/0/7933792e-a2e6-11e4-9c06-00144feab7de.html
3. Barford, P., Canadi, I., Krushevskaja, D., Ma, Q., Muthukrishnan, S.: Adscape: harvesting and analyzing online display ads. In: Proceedings of the 23rd International Conference on World Wide Web, WWW'14, pp. 597–608. ACM, New York (2014)
4. Barocas, S., Nissenbaum, H.: Big data's end run around procedural privacy protections. Commun. ACM **57**(11), 31–33 (2014)
5. Big data in education (2015). https://www.edx.org/course/big-data-education-teacherscollegex-bde1x
6. Big data in government, defense and homeland security 2015–2020 (2015). http://www.prnewswire.com/news-releases/big-data-in-government-defense-and-homeland-security-2015---2020.html
7. Bishop, C.M.: Pattern Recognition and Machine Learning (Information Science and Statistics). Springer, New York (2006)
8. Bork, P., Jensen, L., von Mering, C., Ramani, A., Lee, I., Marcott, E.: Protein interaction networks from yeast to human. Curr. Opin. Struct. Biol. **14**(3), 292–299 (2004)
9. Breiman, L.: Random forests. Mach. Learn. **45**(1), 5–32 (2001)
10. Calders, T., Verwer, S.: Three naive Bayes approaches for discrimination-free classification. Data Min. Knowl. Disc. **21**(2), 277–292 (2010)
11. Cover, T.M., Thomas, J.A.: Elements of Information Theory. Wiley, New York (2012)
12. Datta, A., Tschantz, M., Datta, A.: Automated experiments on ad privacy settings: a tale of opacity, choice, and discrimination. In: Proceedings on Privacy Enhancing Technologies (PoPETs 2015), pp. 92–112 (2015)
13. Datta, A., Datta, A., Procaccia, A., Zick, Y.: Influence in classification via cooperative game theory. In: Proceedings of the 24th International Joint Conference on Artificial Intelligence (IJCAI 2015), pp. 511–517 (2015)
14. Datta, A., Sen, S., Zick, Y.: Algorithmic transparency via quantitative input influence: theory and experiments with learning systems. In: Proceedings of 37th Symposium on Security and Privacy (Oakland 2016), pp. 598–617 (2016)
15. Dwork, C., Hardt, M., Pitassi, T., Reingold, O., Zemel, R.: Fairness through awareness. In: Proceedings of the 3rd Innovations in Theoretical Computer Science Conference (ITCS 2012), pp. 214–226 (2012)
16. E.G. Griggs v. Duke Power Co., 401 U.S. 424, 91 S. Ct. 849, 28 L. Ed. 2d 158 (1977)
17. Feldman, M., Friedler, S.A., Moeller, J., Scheidegger, C., Venkatasubramanian, S.: Certifying and removing disparate impact. In: Proceedings of the 21th ACM SIGKDD International Conference on Knowledge Discovery and Data Mining, KDD'15, pp. 259–268. ACM, New York (2015)
18. Guha, S., Cheng, B., Francis, P.: Challenges in measuring online advertising systems. In: Proceedings of the 10th ACM SIGCOMM Conference on Internet Measurement, IMC'10, pp. 81–87. ACM, New York (2010)
19. Guyon, I., Elisseeff, A.: An introduction to variable and feature selection. J. Mach. Learn. Res. **3**, 1157–1182 (2003)
20. Janzing, D., Balduzzi, D., Grosse-Wentrup, M., Schölkopf, B.: Quantifying causal influences. Ann. Statist. **41**(5), 2324–2358 (2013)
21. Jelveh, Z., Luca, M.: Towards diagnosing accuracy loss in discrimination-aware classification: an application to predictive policing. In: Fairness, Accountability and Transparency in Machine Learning, pp. 137–141 (2014)

22. Kamishima, T., Akaho, S., Sakuma, J.: Fairness-aware learning through regularization approach. In: Proceedings of the 2011 IEEE 11th International Conference on Data Mining Workshops (ICDMW 2011), pp. 643–650 (2011)
23. Keinan, A., Sandbank, B., Hilgetag, C., Meilijson, I., Ruppin, E.: Fair attribution of functional contribution in artificial and biological networks. Neural Comput. **16**(9), 1887–1915 (2004)
24. Lécuyer, M., Ducoffe, G., Lan, F., Papancea, A., Petsios, T., Spahn, R., Chaintreau, A., Geambasu, R.: Xray: enhancing the web's transparency with differential correlation. In: Proceedings of the 23rd USENIX Conference on Security Symposium, SEC'14, pp. 49–64. USENIX Association, Berkeley (2014)
25. Lecuyer, M., Spahn, R., Spiliopolous, Y., Chaintreau, A., Geambasu, R., Hsu, D.: Sunlight: Fine-grained targeting detection at scale with statistical confidence. In: Proceedings of the 22nd ACM SIGSAC Conference on Computer and Communications Security, CCS'15, pp. 554–566. ACM, New York (2015)
26. Letham, B., Rudin, C., McCormick, T.H., Madigan, D.: Interpretable classifiers using rules and Bayesian analysis: building a better stroke prediction model. Ann. Appl. Stat. **9**(3), 1350–1371 (2015)
27. Lichman, M.: UCI machine learning repository (2013). http://archive.ics.uci.edu/ml
28. Lindelauf, R., Hamers, H., Husslage, B.: Cooperative game theoretic centrality analysis of terrorist networks: the cases of Jemaah Islamiyah and Al Qaeda. Eur. J. Oper. Res. **229**(1), 230–238 (2013)
29. Maschler, M., Solan, E., Zamir, S.: Game Theory. Cambridge University Press, Cambridge (2013)
30. Michalak, T., Rahwan, T., Szczepanski, P., Skibski, O., Narayanam, R., Wooldridge, M., Jennings, N.: Computational analysis of connectivity games with applications to the investigation of terrorist networks. In: Proceedings of the 23rd International Joint Conference on Artificial Intelligence (IJCAI 2013), pp. 293–301 (2013)
31. Murdoch, T.B., Detsky, A.S.: The inevitable application of big data to health care. http://jama.jamanetwork.com/article.aspx?articleid=1674245
32. National longitudinal surveys (2017). http://www.bls.gov/nls/
33. O'Donnell, R.: Analysis of Boolean Functions. Cambridge University Press, New York (2014)
34. Perry, W.L., McInnis, B., Price, C.C., Smith, S.C., Hollywood, J.S.: Predictive policing: the role of crime forecasting in law enforcement operations. RAND Corporation, Santa Monica (2013)
35. Podesta, J., Pritzker, P., Moniz, E., Holdern, J., Zients, J.: Big data: seizing opportunities, preserving values. Technical Report, Executive Office of the President - the White House (2014)
36. Rényi, A.: On measures of entropy and information. In: Proceedings of the Fourth Berkeley Symposium on Mathematical Statistics and Probability, Volume 1: Contributions to the Theory of Statistics, pp. 547–561. University of California Press, Berkeley (1961)
37. Rüping, S.: Learning interpretable models. Ph.D. Thesis, Dortmund University of Technology (2006). Http://d-nb.info/997491736
38. Shannon, C.E.: A mathematical theory of communication. Bell Syst. Tech. J. **27**(3), 379–423 (1948)
39. Shapley, L.: A value for *n*-person games. In: Contributions to the Theory of Games, vol. 2, Annals of Mathematics Studies, No. 28, pp. 307–317. Princeton University Press, Princeton (1953)
40. Shapley, L.S., Shubik, M.: A method for evaluating the distribution of power in a committee system. Am. Polit. Sci. Rev. **48**(3), 787–792 (1954)
41. Smith, G.: Quantifying information flow using min-entropy. In: Proceedings of the 8th International Conference on Quantitative Evaluation of Systems (QEST 2011), pp. 159–167 (2011)
42. Strumbelj, E., Kononenko, I.: An efficient explanation of individual classifications using game theory. J. Mach. Learn. Res. **11**, 1–18 (2010)

43. The National Center for Fair and Open Testing: 850+ colleges and universities that do not use SAT/ACT scores to admit substantial numbers of students into bachelor degree programs (2015). http://www.fairtest.org/university/optional
44. Tibshirani, R.: Regression shrinkage and selection via the lasso: a retrospective. J. R. Stat. Soc. Ser. B **73**(3), 273–282 (2011)
45. University, G.W.: Standardized test scores will be optional for GW applicants (2015). https://gwtoday.gwu.edu/standardized-test-scores-will-be-optional-gw-applicants
46. Ustun, B., Tracà, S., Rudin, C.: Supersparse linear integer models for interpretable classification. ArXiv e-prints (2013). http://arxiv.org/pdf/1306.5860v1
47. Young, H.: Monotonic solutions of cooperative games. Int. J. Game Theory **14**(2), 65–72 (1985)
48. Zemel, R., Wu, Y., Swersky, K., Pitassi, T., Dwork, C.: Learning fair representations. In: Proceedings of the 30th International Conference on Machine Learning (ICML 2013), pp. 325–333 (2013)

Learning Interpretable Classification Rules with Boolean Compressed Sensing

Dmitry M. Malioutov, Kush R. Varshney, Amin Emad, and Sanjeeb Dash

Abstract An important problem in the context of supervised machine learning is designing systems which are interpretable by humans. In domains such as law, medicine, and finance that deal with human lives, delegating the decision to a black-box machine-learning model carries significant operational risk, and often legal implications, thus requiring interpretable classifiers. Building on ideas from Boolean compressed sensing, we propose a rule-based classifier which explicitly balances accuracy versus interpretability in a principled optimization formulation. We represent the problem of learning conjunctive clauses or disjunctive clauses as an adaptation of a classical problem from statistics, Boolean group testing, and apply a novel linear programming (LP) relaxation to find solutions. We derive theoretical results for recovering sparse rules which parallel the conditions for exact recovery of sparse signals in the compressed sensing literature. This is an exciting development in interpretable learning where most prior work has focused on heuristic solutions. We also consider a more general class of rule-based classifiers, checklists and scorecards, learned using ideas from threshold group testing. We show competitive classification accuracy using the proposed approach on real-world data sets.

Abbreviations

1Rule: Boolean compressed sensing-based single rule learner
C5.0: C5.0 Release 2.06 algorithm with rule set option in SPSS
CART: Classification and regression trees algorithm in Matlab's classregtree function
CS: Compressed sensing

D.M. Malioutov (✉) • K.R. Varshney • S. Dash
IBM T. J. Watson Research Center, Yorktown Heights, NY, USA
e-mail: dmalioutov@us.ibm.com; krvarshn@us.ibm.com; sanjeebd@us.ibm.com

A. Emad
Institute for Genomic Biology, University of Illinois, Urbana Champaign, Urbana, IL, USA

1218 Thomas M. Siebel Center for Computer Science, University of Illinois, Urbana, IL 61801, USA
e-mail: emad2@illinois.edu

T. Cerquitelli et al. (eds.), *Transparent Data Mining for Big and Small Data*,
Studies in Big Data 32, DOI 10.1007/978-3-319-54024-5_5

DList: Decision lists algorithm
ILPD: Indian liver patient dataset
Ionos: Ionosphere dataset
IP: Integer programming
kNN: The k-nearest neighbor algorithm
Liver: BUPA liver disorders dataset
LP: Linear programming
Parkin: Parkinsons dataset
Pima: Pima Indian diabetes dataset
RuB: Boosting approach rule learner
RuSC: Set covering approach rule learner
SCM: Set covering machine
Sonar: Connectionist bench sonar dataset
SQGT: Semiquantitative group testing
SVM: Support vector machine
TGT: Threshold group testing
Trans: Blood transfusion service center dataset
TrBag: The random forests classifier in Matlab's TreeBagger class
UCI: University of California Irvine
WDBC: Wisconsin diagnostic breast cancer dataset

1 Introduction

A great variety of formulations have been developed for the supervised learning problem, but the most powerful among these, such as kernel support vector machines (SVMs), gradient boosting, random forests, and neural networks are essentially black boxes in the sense that it is difficult for humans to interpret them. In contrast, early heuristic approaches such as decision lists that produce Boolean rule sets [9, 10, 44] and decision trees, which can be distilled into Boolean rule sets [43], have a high level of interpretability and are still widely used by analytics practitioners for this reason despite being less accurate. It has been frequently noted that Boolean rules with a small number of terms are the most well-received, trusted, and adopted outputs by human decision makers [32].

We approach this problem from a new computational lens. The sparse signal recovery problem in the Boolean algebra, now often known as Boolean compressed sensing, has received much recent research interest in the signal processing literature [2, 8, 21, 27, 34, 38, 46]. The problem has close ties to classic nonadaptive group testing [16, 17], and also to the compressed sensing and sparse signal recovery literature [6]. Building upon strong results for convex relaxations in compressed sensing, one advance has been the development of a linear programming (LP) relaxation with exact recovery guarantees for Boolean compressed sensing [34].[1] In this chapter,

[1]Other approaches to approximately solve group testing include greedy methods and loopy belief propagation; see references in [34].

we use the ideas of Boolean compressed sensing to address the problem of learning classification rules based on generalizable Boolean formulas from training data, thus drawing a strong connection between these two heretofore disparate problems. We develop a sparse signal formulation for the supervised learning of Boolean classification rules and develop a solution through LP relaxation.

The primary contribution of this work is showing that the problem of learning sparse conjunctive clause rules and sparse disjunctive clause rules from training samples can be represented as a group testing problem, and that we can apply an LP relaxation that resembles the basis pursuit algorithm to solve it [35]. Despite the fact that learning single clauses is NP-hard, we also establish conditions under which, if the data can be perfectly classified by a sparse Boolean rule, the relaxation recovers it exactly. To the best of our knowledge, this is the first work that combines compressed sensing ideas with classification rule learning to produce optimal sparse (interpretable) rules.

Due to the practical concern of classifier interpretability for adoption and impact [24, 49], there has been a renewed interest in rule learning that attempts to retain the interpretability advantages of rules, but changes the training procedures to be driven by optimizing an objective rather than being heuristic in nature [3, 14, 23, 28, 31, 45]. Set covering machines (SCM) formulate rule learning with an optimization objective similar to ours, but find solutions using a greedy heuristic rather than the LP relaxation that we propose [39]. Local analysis of data [5] also considers an optimization formulation to find compact rules, but they treat positive and negative classes separately and do not explicitly balance errors vs. interpretability. Maximum monomial agreement [19] also uses linear programming for learning rules, but they do not encourage sparsity. Two new interpretable rule learning methods which have substantially more complicated optimization formulations appeared after the initial presentation of this work [48, 50, 54].

In addition to considering ordinary classification rules, we show that the connection to sparse recovery can shed light on a related problem of learning checklists and scorecards, which are widely used in medicine, finance and insurance as a simple rubric to quickly make decisions. Such scorecards are typically constructed manually based on domain expert intuition. We show that the problem of automatically learning checklists from data can also be viewed as a version of Boolean sparse recovery, with strong connections to the threshold group-testing problem.

In today's age of big data, machine learning algorithms are often trained in the presence of a large number of features and a large number of samples. In our proposed LP formulation, this results in a large number of variables and a large number of constraints, i.e., columns and rows in the sensing matrix (using the terminology of compressed sensing). We would like to be able to reduce the number of columns and rows before solving the LP for tractability.

The certifiable removal of variables that will not appear in the optimal solution is known as *screening* [11, 20, 33, 51–53, 55–57]; in this chapter, we develop novel screening tests for Boolean compressed sensing and nonadaptive group testing [12].

Specifically, we develop two classes of screening tests: simple screening rules that arise from misclassification error counting arguments, and rules based on obtaining a feasible primal-dual pair of LP solutions.

Additionally, we develop a novel approach to reduce the number of rows [13]. In a sequential setting, we can certify reaching a near-optimal solution while only solving the LP on a small fraction of the available samples. Related work has considered progressive cross-validation error and argues that when this error 'stabilizes' while being exposed to a stream of i.i.d. training samples, then the classifier reaches near-optimal test classification accuracy [4]. The *learning curve* literature in machine learning has considered how the generalization error evolves as a function of the received number of samples [29, 40, 42]. In the context of ordinary (non-Boolean) compressed sensing, [36] has developed sequential tests to establish that a sufficient number of measurements has been obtained to recover the correct sparse signal.

We demonstrate the proposed approach on several real-world data sets and find that the proposed approach has better accuracy than heuristic decision lists and has similar interpretability. The accuracy is in fact on par with less interpretable weighted rule set induction, and not far off from the best non-interpretable classifiers. We find that our screening tests are able to safely eliminate a large fraction of columns, resulting in large decreases in running time. We also find that through our row sampling approach, we are able to obtain accurate near-optimal interpretable classification rules while training on only a small subset of the samples.

The remainder of this chapter is organized as follows. In Sect. 2 we provide a brief review of group testing and formulate supervised binary classification as a group testing problem. In Sect. 3, we present an LP relaxation for rule learning and show that it can recover rules exactly under certain conditions. Section 4 extends the discussion to scorecards. Section 5 develops screenings tests and row sampling. Empirical evaluations on real-world data are given in Sect. 6. We conclude with a summary and discussion in Sect. 7.

2 Boolean Rule Learning as Group Testing

In this section, we formulate the problem of learning sparse AND-clauses such as:

$$\text{height} \leq 6 \text{ feet AND } \text{weight} > 200 \text{ pounds}$$

and OR-clauses such as:

$$\text{Smoke} = \text{True OR } \text{Exercise} = \text{False OR } \text{Blood Pressure} = \text{High}$$

from training data via group testing.

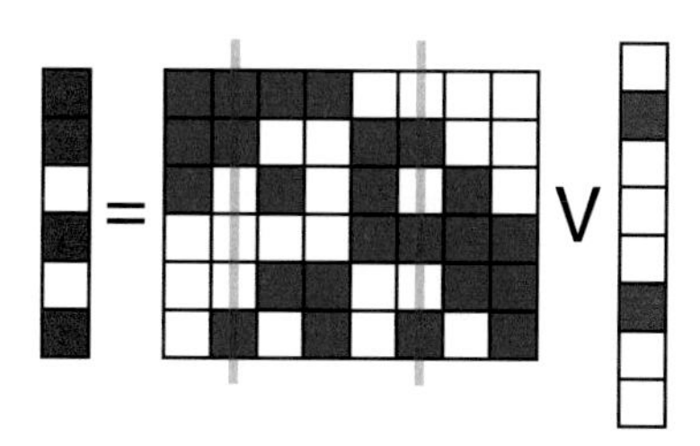

Fig. 1 Illustration of group testing, $\mathbf{y} = \mathbf{A} \vee \mathbf{w}$: measurements $\mathbf{y}$ are Boolean combinations of the sparse unknown vector $\mathbf{w}$. Matrix $\mathbf{A}$ specifies which subjects participate in which pooled test

2.1 The Group Testing Problem

Group testing [16] was developed during World War II, when the US military needed to conduct expensive medical tests on a large number of soldiers. The key idea was that if the test was performed on a group of soldiers simultaneously, by combining their blood samples into a pooled test, the cost could be dramatically reduced. Consider an $m \times n$ Boolean matrix $\mathbf{A}$, where the rows represent different pools (subsets of subjects) and the columns represent the subjects. An entry a_{ij} is one if subject j is part of a pool i and zero otherwise. The true states of the subjects (unknown when conducting the tests) are represented by vector $\mathbf{w} \in \{0, 1\}^n$. Group testing, in which the result of a test is the OR of all subjects in a pool, results in a Boolean vector $\mathbf{y} \in \{0, 1\}^m$. We summarize the result of all m tests using:

$$\mathbf{y} = \mathbf{A} \vee \mathbf{w}, \tag{1}$$

which represents Boolean matrix-vector multiplication, i.e.,

$$y_i = \bigvee_{j=1}^{n} a_{ij} \wedge w_j. \tag{2}$$

We illustrate this idea in Fig. 1. In the presence of measurement errors,

$$\mathbf{y} = (\mathbf{A} \vee \mathbf{w}) \oplus \mathbf{n}, \tag{3}$$

where $\oplus$ is the XOR operator and $\mathbf{n}$ is a noise vector.

Once the tests have been conducted, the objective is to recover $\mathbf{w}$ from $\mathbf{A}$ and the measured $\mathbf{y}$. The recovery can be stated through the following combinatorial optimization problem:

$$\min \|\mathbf{w}\|_0 \quad \text{such that } \mathbf{y} = \mathbf{A} \vee \mathbf{w}, \mathbf{w} \in \{0, 1\}^n. \tag{4}$$

In the presence of noise we use parameter λ to balance sparsity of $\mathbf{w}$ and the errors $\mathbf{n}$:

$$\min_{\mathbf{w}} \ \lambda \|\mathbf{w}\|_0 + \sum n_i \quad \text{such that } \mathbf{y} = (\mathbf{A} \vee \mathbf{w}) \oplus \mathbf{n}, \quad \mathbf{w}, \mathbf{n} \in \{0, 1\}^n. \tag{5}$$

2.2 Supervised Classification Rule Formulation

We have described group testing; now we show how the formulation can be adapted to rule-based classification. The problem setup of interest is standard binary supervised classification. We are given m labeled training samples $\{(\mathbf{x}_1, y_1), \ldots, (\mathbf{x}_m, y_m)\}$ where the $\mathbf{x}_i \in \mathscr{X}$ are the features in some discrete or continuous space $\mathscr{X}$ and the $y_i \in \{0, 1\}$ are the Boolean labels. We would like to learn a function $\hat{y}(\cdot) : \mathscr{X} \rightarrow \{0, 1\}$ that will accurately generalize to classify unseen, unlabeled feature vectors drawn from the same distribution as the training samples.

In rule-based classifiers, the clauses are made up of individual Boolean terms, e.g., 'weight $>$ 200.' Such a term can be represented by a function $a(\mathbf{x})$ mapping the feature vector to a boolean number. To represent the full diversity and dimensions of the feature space $\mathscr{X}$, we have many such Boolean terms $a_j(\cdot) : \mathscr{X} \rightarrow \{0, 1\}$, $j = 1, \ldots, n$. For each continuous dimension of $\mathscr{X}$, these terms may be comparisons to several suitably chosen thresholds. Then for each of the training samples, we can calculate the truth value for each of the terms, leading to an $m \times n$ truth table $\mathbf{A}$ with entries $a_{ij} = a_j(\mathbf{x}_i)$.

Writing the true labels of the training set as a vector $\mathbf{y}$, we can write the same expression in the classification problem as in group testing (3): $\mathbf{y} = (\mathbf{A} \vee \mathbf{w}) \oplus \mathbf{n}$. In the classification problem, $\mathbf{w}$ is the binary vector to be learned that indicates the rule. The nonzero coefficients directly specify a Boolean clause classification rule which can be applied to new unseen data. This clause is a disjunctive OR-rule. In most of the rule-based classification literature, however, the learning of AND-clauses is preferred. This is easy to handle using DeMorgan's law. If we complement $\mathbf{y}$ and $\mathbf{A}$ prior to the learning, then we have:

$$\mathbf{y} = \mathbf{A} \wedge \mathbf{w} \quad \Leftrightarrow \quad \mathbf{y}^C = \mathbf{A}^C \vee \mathbf{w}. \tag{6}$$

Hence, our results apply to both OR-rules and AND-rules; we focus on the conjunctive case for the remainder of the chapter.

For interpretability and generalization, we are specifically interested in compact Boolean rules, i.e., we would like $\mathbf{w}$ to be sparse, having few non-zero entries. Therefore, the optimization problem to be solved is the same as for group testing (4). We describe our proposed solution next.

3 LP Relaxation

The group testing problem appears closely related to the compressed sensing (CS) problem from the signal processing literature [6]. Both group testing and compressed sensing involve sparse signal recovery, but group testing uses Boolean algebra instead of the typical real-valued linear algebra encountered in CS. In this section, based on their close connection, we show that suitably modified, efficient LP relaxations from compressed sensing can be used to solve the group testing problem, and hence also the classification rule learning problem.

3.1 Boolean Compressed Sensing-Based Formulation

Compressed sensing attempts to find an unknown high-dimensional but sparse real-valued vector $\mathbf{w}$ from a small collection of random measurements $\mathbf{y} = \mathbf{A}\mathbf{w}$, where $\mathbf{A}$ is a random matrix (e.g., with i.i.d. Gaussian entries). The problem is to find the sparsest solution, $\min \|\mathbf{w}\|_0$ such that $\mathbf{y} = \mathbf{A}\mathbf{w}$. It looks very close to group testing (4), except that $\mathbf{y}, \mathbf{A}$, and $\mathbf{w}$ are real-valued, and $\mathbf{A}\mathbf{w}$ denotes the standard matrix-vector product in linear algebra. In compressed sensing, the most popular technique for getting around the combinatorial ℓ_0 objective is to relax it using the convex ℓ_1-norm. This relaxation, known as basis pursuit, results in the following optimization problem:

$$\min \|\mathbf{w}\|_1 \quad \text{such that } \mathbf{y} = \mathbf{A}\mathbf{w}, \tag{7}$$

where $\mathbf{y}$, $\mathbf{w}$, and $\mathbf{A}$ are all real-valued and the product $\mathbf{A}\mathbf{w}$ is the standard matrix-vector product. This optimization problem (7) is a linear program and can be solved efficiently. It has been shown that under certain conditions on the matrix $\mathbf{A}$ and sparsity of $\mathbf{w}$, the ℓ_0 solution and the ℓ_1 solution are equivalent. The work of [34] extends the basis pursuit idea to Boolean algebras.

The challenge in compressed sensing is with the combinatorial nature of the ℓ_0 objective. Additionally, in the Boolean setting, Eq. (1) is not a set of linear constraints. However, if a vector $\mathbf{w}$ satisfies the constraint that $\mathbf{y} = \mathbf{A} \vee \mathbf{w}$, then it also satisfies the pair of ordinary linear inequalities $\mathbf{A}_{\mathscr{P}}\mathbf{w} \geq \mathbf{1}$ and $\mathbf{A}_{\mathscr{Z}}\mathbf{w} = \mathbf{0}$, where $\mathscr{P} = \{i|y_i = 1\}$ is the set of positive tests, $\mathscr{Z} = \{i|y_i = 0\}$ is the set of negative (or zero) tests, and $\mathbf{A}_{\mathscr{P}}$ and $\mathbf{A}_{\mathscr{Z}}$ are the corresponding subsets of rows of $\mathbf{A}$. We refer to the jth column of $\mathbf{A}, \mathbf{A}_{\mathscr{P}}$ and $\mathbf{A}_{\mathscr{Z}}$ as $\mathbf{a}^j, \mathbf{a}^j_{\mathscr{P}}$ and $\mathbf{a}^j_{\mathscr{Z}}$, respectively. The vectors $\mathbf{1}$ and $\mathbf{0}$ are all ones and all zeroes, respectively. These constraints can be incorporated into an LP. Thus the Boolean ℓ_1 problem is the integer program (IP):

$$\begin{aligned} \min \quad & \sum_{j=1}^{n} w_j \\ \text{s.t.} \quad & w_j \in \{0, 1\}, j = 1, \ldots, n \\ & \mathbf{A}_{\mathscr{P}}\mathbf{w} \geq \mathbf{1} \\ & \mathbf{A}_{\mathscr{Z}}\mathbf{w} = \mathbf{0}. \end{aligned} \tag{8}$$

Because of the Boolean integer constraint on the weights, the problem (8) is NP-hard. We can further relax the optimization to the following tractable LP[2]:

[2]Instead of using LP, one can find solutions greedily, as is done in the SCM, which gives a $\log(m)$ approximation. The same guarantee holds for LP with randomized rounding. Empirically, LP tends to find sparser solutions.

$$\min \sum_{j=1}^{n} w_j \tag{9}$$
$$\text{s.t.} \quad 0 \leq w_j \leq 1, j = 1, \ldots, n$$
$$\mathbf{A}_{\mathscr{P}}\mathbf{w} \geq \mathbf{1}$$
$$\mathbf{A}_{\mathscr{Z}}\mathbf{w} = \mathbf{0}.$$

If non-integer w_j are found, we either simply set them to one, or use randomized rounding. An exact solution to the integer LP can be obtained by branch and bound.[3]

Slack variables may be introduced in the presence of errors, when there may not be any sparse rules producing the labels $\mathbf{y}$ exactly, but there are sparse rules that approximate $\mathbf{y}$ very closely. This is the typical case in the supervised classification problem. The LP is then:

$$\min \quad \lambda \sum_{j=1}^{n} w_j + \sum_{i=1}^{m} \xi_i \tag{10}$$
$$\text{s.t.} \quad 0 \leq w_j \leq 1, j = 1, \ldots, n$$
$$\mathbf{A}_{\mathscr{Z}}\mathbf{w} = \boldsymbol{\xi}_{\mathscr{Z}}, \qquad 0 \leq \xi_i, \; i \in \mathscr{Z}.$$
$$\mathbf{A}_{\mathscr{P}}\mathbf{w} + \boldsymbol{\xi}_{\mathscr{P}} \geq \mathbf{1}, \quad 0 \leq \xi_i \leq 1, \; i \in \mathscr{P}$$

The regularization parameter λ trades training error and the sparsity of $\mathbf{w}$.

3.2 Recovery Guarantees

We now use tools from combinatorial group testing [16, 17] to establish results for exact recovery and recovery with small error probability in AND-clause learning via LP relaxation. First, we introduce definitions from group testing.

Definition 1 A matrix $\mathbf{A}$ is *K-separating* if Boolean sums of sets of K columns are all distinct.

Definition 2 A matrix $\mathbf{A}$ is *K-disjunct* if the union (boolean sum) of any K columns does not contain any other column.

Any K-disjunct matrix is also K-separating. The K-separating property for $\mathbf{A}$ is sufficient to allow exact recovery of $\mathbf{w}$ with up to K nonzero entries, but in general, requires searching over all K-subsets out of n [16]. The property of K-

[3]Surprisingly, for many practical datasets the LP formulation obtains integral solutions, or requires a small number of branch and bound steps.

disjunctness, which can be viewed as a Boolean analog of spark [15], is a more restrictive condition that allows a dramatic simplification of the search: a simple algorithm that considers rows where $y_i = 0$ and sets all w_j where $a_{ij} = 1$ to zero and the remaining w_j to one, is guaranteed to recover the correct solution. For non-disjunct matrices this simple algorithm finds feasible but suboptimal solutions.

We recall a lemma on exact recovery by the LP relaxation.

Lemma 1 ([34]) *Suppose there exists a* $\mathbf{w}^*$ *with* K *nonzero entries and* $\mathbf{y} = \mathbf{A} \vee \mathbf{w}^*$. *If the matrix* $\mathbf{A}$ *is* K*-disjunct, then LP solution* $\hat{\mathbf{w}}$ *in* (9) *recovers* $\mathbf{w}^*$, *i.e.,* $\hat{\mathbf{w}} = \mathbf{w}^*$.

This lemma was presented in the group testing context. To apply it to rule learning, we start with classification problems with binary features in which case, the matrix $\mathbf{A}$ simply contains the feature values.[4] A simple corollary of Lemma 1 is that if $\mathbf{A}$ is K-disjunct and there is an underlying error-free K-term AND-rule, then we can recover the rule exactly via (9).

A critical question is when can we expect our features to yield a K-disjunct matrix?

Lemma 2 *Suppose that for each subset of* $K + 1$ *features, we find at least one example of each one of the* 2^{K+1} *possible binary* $(K + 1)$*-patterns among our* m *samples. Then the matrix* $\mathbf{A}$ *is* K*-disjunct.*

Proof Note that there are 2^{K+1} possible binary patterns for K features. Suppose that on the contrary the matrix is not K-disjunct. Without loss of generality, K-disjunctness fails for the first K columns covering the $(K + 1)$-st one. Namely, columns $\mathbf{a}_1, \ldots, \mathbf{a}_{K+1}$ satisfy $\mathbf{a}_{K+1} \subset \cup_{k=1}^{K} \mathbf{a}_k$. This is clearly impossible, since by our assumption the pattern $(0, 0, \ldots, 0, 1)$ for our $K + 1$ variables is among our m samples. □

To interpret the lemma: if features are not strongly correlated, then for any fixed K, for large enough m we will eventually obtain all possible binary patterns. Using a simple union bound, for the case of uncorrelated equiprobable binary features, the probability that at least one of the K-subsets exhibits a non-represented pattern is bounded above by $\binom{n}{K} 2^K (1 - (1/2)^K)^m$. Clearly as $m \to \infty$ this bound approaches zero: with enough samples $\mathbf{A}$ is K-disjunct.

These results also carry over to approximate disjunctness [35] (also known as a weakly-separating design) to develop less restrictive conditions when we allow a small probability of recovery error [37, 41].

In the case of classification with continuous features, we discretize feature dimension x_j using thresholds $\theta_{j,1} \leq \theta_{j,2} \leq \cdots \leq \theta_{j,D}$ such that the columns of $\mathbf{A}$ corresponding to x_j are the outputs of Boolean indicator functions $I_{x_j \leq \theta_{j,1}}(\mathbf{x}), \ldots, I_{x_j \leq \theta_{j,D}}(\mathbf{x}), I_{x_j > \theta_{j,1}}(\mathbf{x}), \ldots, I_{x_j > \theta_{j,D}}(\mathbf{x})$. This matrix is not disjunct

[4] In general it will contain the features and their complements as columns. However, with enough data, one of the two choices will be removed by zero-row elimination beforehand.

because, e.g., $I_{x_j>\theta_{j,1}}(\mathbf{x}) \geq I_{x_j>\theta_{j,2}}(\mathbf{x})$. However, without loss of generality, for each feature we can remove all but one of the corresponding columns of $\mathbf{A}$ as discussed in Sect. 5.1. Through this reduction we are left with a simple classification problem with binary features; hence the results in Lemma 1 apply to continuous features.

3.3 Multi-Category Classification

Extending the proposed rule learner from binary classification to M-ary classification is straightforward through one-vs-all and all-vs-all constructions, as well as a Venn diagram-style approach in which we expand the $\mathbf{y}$ and $\mathbf{w}$ vectors to be matrices $\mathbf{Y} = \left[\mathbf{y}^1 \cdots \mathbf{y}^{\lceil \log_2 M \rceil}\right]$ and $\mathbf{W} = \left[\mathbf{w}^1 \cdots \mathbf{w}^{\lceil \log_2 M \rceil}\right]$. We briefly explain how this can be done for $M = 4$ classes with labels $c_1, \ldots, c_4$. We first construct the label matrix with value 1 in $\mathbf{y}^1$ if a sample has label c_1 or c_2 and zero otherwise. Similarly, an element of $\mathbf{y}^2$ takes value 1 if a sample has label c_1 or c_3 and zero otherwise. The problem of interest then becomes $\mathbf{Y} \approx \mathbf{A} \vee \mathbf{W}$. The constraints in the LP relaxation become $\mathbf{A}_{\mathscr{P}_1}\mathbf{w}^1 \geq \mathbf{1}$, $\mathbf{A}_{\mathscr{P}_2}\mathbf{w}^2 \geq \mathbf{1}$, $\mathbf{A}_{\mathscr{Z}_1}\mathbf{w}^1 = \mathbf{0}$, and $\mathbf{A}_{\mathscr{Z}_2}\mathbf{w}^2 = \mathbf{0}$, where $\mathscr{P}_1 = \{i|y_{1,i} = 1\}$, $\mathscr{P}_2 = \{i|y_{2,i} = 1\}$, $\mathscr{Z}_1 = \{i|y_{1,i} = 0\}$, and $\mathscr{Z}_2 = \{i|y_{2,i} = 0\}$. From a solution $\mathbf{w}^1, \mathbf{w}^2$, we will have two corresponding AND-rules which we denote r_1, r_2. A new sample is classified as c_1 if $r_1 \wedge r_2$, as c_2 if $r_1 \wedge \neg r_2$, as c_3 if $\neg r_1 \wedge r_2$, and as c_4 if $\neg r_1 \wedge \neg r_2$.

4 Learning Scorecards Using Threshold Group Testing

We now build upon our approach to learn sparse Boolean AND or OR rules and develop a method for learning interpretable scorecards using sparse signal representation techniques. In the face of high complexity, checklists and other simple scorecards can significantly improve people's performance on decision-making tasks [26]. An example of such a tool in medicine, the *clinical prediction rule*, is a simple decision-making rubric that helps physicians estimate the likelihood of a patient having or developing a particular condition in the future [1]. An example of a clinical prediction rule for estimating the risk of stroke, known as the CHADS$_2$ score, is shown in Table 1 [25]. The health worker determines which of the five diagnostic indicators a patient exhibits and adds the corresponding points together. The higher the total point value is, the greater the likelihood the patient will develop a stroke. This rule was manually crafted by health workers and notably contains few conditions with small integer point values and is extremely interpretable by people.

Recent machine learning research has attempted to learn clinical prediction rules that generalize accurately from large-scale electronic health record data rather than relying on manual development [31, 48]. The key aspect of the problem is maintaining the simplicity and interpretability of the learned rule to be similar to

Table 1 $CHADS_2$ clinical prediction rule for estimating risk of stroke

Condition	Points
Congestive heart failure	1
Hypertension	1
Age $\geq$ 75	1
Diabetes mellitus	1
Prior stroke, transient ischemic attack, or thromboembolism	2

the hand-crafted version, in order to enable trust and adoption by users in the health care industry. We again employ sparsity as a proxy for interpretability of the rules.

In the previous sections, the form of the classifier we considered was a sparse AND-rule or OR-rule whereas here, we would like to find a sparse set of conditions or features with small integer coefficients that are added together to produce a score. Such a model is between the "1-of-N" and "N-of-N" forms implied by OR-rules and AND-rules, with N active variables. With unit weights on the columns of A it can be viewed as an M-of-N rule. For learning Boolean rules, our Boolean compressed sensing formulation had a close connection with the group testing problem whereas here, the connection is to the *semiquantitative* group testing (SQGT) problem, or more precisely to its special case of threshold group testing (TGT) [21].

4.1 Threshold Group Testing

We start by describing the TGT model and then show how to formulate the interpretable rule learning problem as TGT. Let n, m, and d denote the total number of subjects, the number of tests, and the number of defectives, respectively. As we defined in Sect. 2.1, $\mathbf{A} \in \{0, 1\}^{m \times n}$ is a binary matrix representing the assignment of subjects to each test, and $\mathbf{y} \in \{0, 1\}^m$ is the binary vector representing the error-free results of the tests. Let $\mathscr{D}_t$ be the true set of defectives and binary vector $\mathbf{w}_t \in \{0, 1\}^n$ represent which subject is a defective. In the TGT model, one has

$$\mathbf{y} = f_\eta(\mathbf{A}\mathbf{w}_t), \tag{11}$$

where $f_\eta(\cdot)$ is a quantizing function with threshold η, such that $f_\eta(x) = 0$ if $x < \eta$ and $f_\eta(x) = 1$ if $x \geq \eta$. The goal is to recover the unknown vector $\mathbf{w}$ given the test matrix $\mathbf{A}$ and the vector of test results $\mathbf{y}$.

4.2 LP Formulation for Threshold Group Testing

First, we start by considering the simplest model in which there are no errors in the vector of labels, and the threshold η is known in advance. In this case, and given that $\mathbf{w}_t$ is a sparse vector, one can find the sparsest binary vector that satisfies (11). However, this combinatorial problem is not computationally feasible for large datasets.

To overcome this problem, we use a similar relaxation to the one used in Sect. 3.1. We note that a vector $\mathbf{w}$ that satisfies the constraints imposed by the TGT model in (11), must also satisfy the pair of ordinary linear inequalities:

$$\mathbf{A}_{\mathscr{P}}\mathbf{w} \geq \eta \mathbf{1}, \tag{12}$$

$$\mathbf{A}_{\mathscr{Z}}\mathbf{w} < \eta \mathbf{1}, \tag{13}$$

where, same as before, $\mathscr{P} = \{i|\mathbf{y}(i) = 1\}$ and $\mathscr{Z} = \{i|\mathbf{y}(i) = 0\}$ are the sets of positive and negative tests. Using a convex l_1-norm instead of $||\mathbf{w}||_0$, and relaxing the binary constraint we obtain

$$\min \; ||\mathbf{w}||_1 \tag{14}$$

$$\text{s.t.} \; 0 \leq \mathbf{w}(j) \leq 1, \quad j = 1, \ldots, n \tag{15}$$

$$\mathbf{A}_{\mathscr{P}}\mathbf{w} \geq \eta \mathbf{1}, \tag{16}$$

$$\mathbf{A}_{\mathscr{Z}}\mathbf{w} \leq (\eta - 1)\mathbf{1}, \tag{17}$$

In presence of noise, we introduce slack variables ξ to allow violation of a small subset of the constraints. We also allow the threshold η to not be known a priori, and learn it from data. We propose the following formulation to jointly find the threshold η and the defective items (or our interpretable rules):

$$\begin{aligned} \min \; & ||\mathbf{w}||_1 + \lambda ||\boldsymbol{\xi}||_1 \\ \text{s.t.} \; & 0 \leq \mathbf{w}(j) \leq 1, \quad j = 1, \ldots, n \\ & 0 \leq \boldsymbol{\xi}(i) \leq 1, \quad i \in \mathscr{P} \\ & 0 \leq \boldsymbol{\xi}(i), \quad i \in \mathscr{Z} \\ & \mathbf{A}_{\mathscr{P}}\mathbf{w} + \boldsymbol{\xi}_{\mathscr{P}} \geq \eta \mathbf{1}, \\ & \mathbf{A}_{\mathscr{Z}}\mathbf{w} \leq \eta \mathbf{1}, \\ & 0 \leq \eta \leq n. \end{aligned} \tag{18}$$

4.3 Learning Scorecards with Threshold Group Testing

Our goal is to learn an interpretable function $\hat{y}(\cdot) : \mathscr{X} \rightarrow \{0, 1\}$, given m labeled training samples $\{(\mathbf{x}_1, y_1), \ldots, (\mathbf{x}_m, y_m)\}$. To formulate this problem as TGT, we form the vector of test results according to $\mathbf{y}(i) = y_i$, $i = 1, 2, \ldots, m$; also, we form the test matrix $\mathbf{A}$ according to $\mathbf{A}(i, j) = a_j(\mathbf{x}_i)$, where $a_j(\cdot) : \mathscr{X} \rightarrow \{0, 1\}$, $j = 1, \ldots, n$, are simple Boolean terms (e.g. age $\geq$ 75). Furthermore, we assume that at most d simple Boolean terms govern the relationship between the labels and the features, i.e. $|\mathscr{D}_t| \leq d$, and this sparse set of terms are encoded in the unknown sparse vector $\mathbf{w}_t$. In addition, we assume that this relationship has the form of a "M-of-N" rule table; in other words, $y_i = 1$ if at least M terms of N are satisfied and $y_i = 0$, otherwise. Therefore, by setting $\eta = M$ and $d = N$, we can write this relationship as (11). Consequently, in order to find the set of interpretable rules corresponding to a "M-of-N" rule table, we need to recover the sparse vector $\mathbf{w}_t$ given $\mathbf{A}$ and $\mathbf{y}$.

4.4 Theoretical Guarantees for TGT

We now summarize results on recovery of sparse rules in the threshold group testing formulation, which generalize our results in Sect. 3.2. We start by defining a generalization of binary d-disjunct codes [30] which is studied under different names in the literature such as cover-free families (e.g. see [7, 18, 47]).

Definition 1 A matrix $\mathbf{A} \in [2]^{m \times n}$ is a (d, η)-disjunct matrix if for any two disjoint sets of column-indices $\mathscr{C}_z$ and $\mathscr{C}_o$,[5] where $|\mathscr{C}_z| = d - \eta + 1$, $|\mathscr{C}_o| = \eta$, and $\mathscr{C}_o \cap \mathscr{C}_z = \varnothing$, there exists at least one row indexed by r such that

$$\mathbf{A}(r, j) = 1 \quad \forall j \in \mathscr{C}_o,$$

$$\mathbf{A}(r, j) = 0 \quad \forall j \in \mathscr{C}_z.$$

Using (d, η)-disjunctness, the following theorem can be established using results in [21].

Theorem 1 *Let $\mathbf{A}$ be a (d, η)-disjunct binary matrix. The LP formulation* (14)–(17) *will uniquely identify the true set of defectives, i.e. $\hat{\mathbf{w}} = \mathbf{w}_t$, as long as $\eta \leq |\mathscr{D}_t| \leq d$.*

The main idea in proving this theorem is that we introduce a reversible transformation that converts the TGT model into another model resembling the Boolean compressed sensing formulation. Given this new formulation, we prove that the LP relaxation can uniquely identify the defectives, hence recovering the sparse set of rules.

[5]Here, the subscript "z" stands for zero and "o" stands for one.

5 Screening and Row Sampling

The formulation (10) produces an LP that becomes challenging for data sets with large numbers of features and large numbers of thresholds on continuous features. Our aim in this section is to provide computationally inexpensive pre-computations which allow us to eliminate the majority of the columns in the $\mathbf{A}$ matrix by providing a certificate that they cannot be part of the optimal solution to the Boolean ℓ_1-IP in (8). The LP also becomes challenging for large numbers of training samples. We develop an approach to only solve the LP for a subset of rows of $\mathbf{A}$ and obtain a near-optimal solution to the full problem.

We first discuss screening approaches that simply examine the nonzero patterns of $\mathbf{A}$ and $\mathbf{y}$, and then discuss screening tests that use a feasible primal-dual pair for the LP to screen even more aggressively. The tests can be applied separately or sequentially; in the end, it is the union of the screened columns that is of importance. Finally, we put forth a row sampling approach also based on feasible primal-dual pairs.

5.1 *Simple Screening Tests*

We first observe that a positive entry in column $\mathbf{a}^j_{\mathscr{Z}}$ corresponds to a false alarm error if the column is active (i.e., if the corresponding $w_j = 1$). The potential benefit of including column j is upper bounded by the number of positive entries in $\mathbf{a}^j_{\mathscr{P}}$. The first screening test is simply to remove columns in which $\|\mathbf{a}^j_{\mathscr{Z}}\|_0 \geq \|\mathbf{a}^j_{\mathscr{P}}\|_0$.

An additional test compares pairs of columns j and j' for different threshold values of the same continuous feature dimension of $\mathscr{X}$. We note that such columns form nested subsets in the sense of sets of nonzero entries. If θ_j and $\theta_{j'}$ are the thresholds defining $a_j(\cdot)$ and $a_{j'}(\cdot)$ with $\theta_j < \theta_{j'}$, then $\{k \mid x_k < \theta_j\} \subset \{k \mid x_k < \theta_{j'}\}$. Looking at the difference in the number of positive entries between columns of $\mathbf{A}_{\mathscr{Z}}$ and the difference in the number of positive entries between columns of $\mathbf{A}_{\mathscr{P}}$, we never select column j instead of column j' if $\|\mathbf{a}^{j'}_{\mathscr{P}}\|_0 - \|\mathbf{a}^j_{\mathscr{P}}\|_0 > \|\mathbf{a}^{j'}_{\mathscr{Z}}\|_0 - \|\mathbf{a}^j_{\mathscr{Z}}\|_0$ by a similar argument as before.

We consider two variations of this pairwise relative cost-redundancy test: first, only comparing pairs of columns such that $j' = j + 1$ when the columns are arranged by sorted threshold values, and second, comparing all pairs of columns for the same continuous feature, which has higher computational cost but can yield a greater fraction of columns screened. Although most applicable to columns corresponding to different threshold values of the same continuous feature of $\mathscr{X}$, the same test can be conducted for any two columns j and j' across different features.

5.2 Screening Tests Based on a Feasible Primal-Dual Pair

We also introduce another kind of screening tests based on LP duality theory, which can further reduce the number of columns when a primal-dual feasible pair is available. We describe a cost-effective way to provide such primal dual pairs. Specifically, we first reformulate (10) along with the requirement that $\mathbf{w}$ is a Boolean vector as a minimum weight set cover problem. Then, if we have a feasible binary primal solution available, we can produce certificates that w_j cannot be nonzero in the optimal solution as follows. If by setting $w_j = 1$ and recomputing the dual, the dual objective function value exceeds the primal objective function value, then any solution with $w_j = 1$ is strictly inferior to the feasible binary primal solution that we started with and we can remove column $\mathbf{a}^j$. Thus a key step is finding feasible binary primal and dual solutions on which we can base the screening. Note that this test explicitly assumes that we want integral solutions to (10); the columns removed would not be present in an optimal binary solution, but could be present in an optimal fractional solution. For the sake of readability, we postpone the detailed derivations of LP-duality based screening tests to Appendix 2.

5.3 Row Sampling

The previous section was concerned with removing columns from $\mathbf{A}$ whereas this section is concerned with removing rows. Suppose that we have a large number $\bar{m}$ of samples available, and we believe that we can learn a near-optimal interpretable classifier from a much smaller subset of $m \ll \bar{m}$ samples. We proceed to develop a certificate which shows that when m is large enough, the solution of (10) on the smaller subset achieves a near optimal solution on the full data set.

To compare the solutions of LPs defined with a different number of samples, we compare their "scaled" optimal objective values, i.e., we divide the objective value by the number of samples (which is equal to m, the number of rows in $\mathbf{A}$). Therefore, we compute and compare error rates rather than raw errors. Let $(\hat{\mathbf{w}}^m, \boldsymbol{\xi}^m)$ and $(\hat{\mathbf{w}}^{\bar{m}}, \boldsymbol{\xi}^{\bar{m}})$ be the optimal solutions for the small LP with m samples and large LP with $\bar{m}$ samples, respectively, with corresponding scaled optimal objective values f_m and $f_{\bar{m}}$. Also, matrices corresponding to the small LP are $\mathbf{A}$, $\mathbf{A}_{\mathscr{P}}$, $\mathbf{A}_{\mathscr{Z}}$ and to the large LP are $\bar{\mathbf{A}}$, $\bar{\mathbf{A}}_{\mathscr{P}}$, $\bar{\mathbf{A}}_{\mathscr{Z}}$. The first m rows of $\bar{\mathbf{A}}$ are $\mathbf{A}$ and the first p entries of $\bar{\mathbf{A}}_{\mathscr{P}}$ are $\mathbf{A}_{\mathscr{P}}$. By definition $\mathbf{A}$ is a submatrix of $\bar{\mathbf{A}}$. Therefore, $f_m \to f_{\bar{m}}$ as $m \to \bar{m}$ and we would like to bound $|f_m - f_{\bar{m}}|$ without solving the large LP.

We show how to extend the primal and the dual solutions of the small LP and obtain both a lower and an upper bound on the scaled optimal objective value of the large LP. To create a feasible primal solution for the large LP we can extend the vector $\hat{\mathbf{w}}^m$ from the small LP by computing the associated errors on the large LP: $\boldsymbol{\xi}^{\bar{m}}_{\mathscr{Z}} = \mathbf{A}_{\mathscr{Z}}\hat{\mathbf{w}}^m$ and

$$\boldsymbol{\xi}_{\mathscr{P}}^{\bar{m}} = \begin{cases} 0 & \text{if } \mathbf{A}_{\mathscr{P}} \hat{\mathbf{w}}^m \geq 1 \\ 1 & \text{otherwise.} \end{cases}$$

This pair $(\hat{\mathbf{w}}^m, \boldsymbol{\xi}^{\bar{m}})$ is feasible for the large LP and the scaled objective value provides an upper bound on $f_{\bar{m}}$. In a similar manner, the solution to the small IP can be extended to a feasible solution of the large IP, thereby giving an upper bound to the optimal IP solution; note that $f_{\bar{m}}$ gives a lower bound.

To find a lower bound on $f_{\bar{m}}$ we extend the dual solution of the small LP to give a feasible (but generally sub-optimal) dual solution of the large LP. We describe the details in Appendix 3.

The discussion so far has been in the batch setting where all training samples are available at the outset; the only goal is to reduce the computations in solving the linear program. We may also be in an online setting where we can request additional i.i.d. samples and would like to declare that we are close to a solution that will not change much with additional samples. This may be accomplished by computing expected upper and lower bounds on the objective value of the large LP as described in [13].

6 Empirical Results

In this section, we evaluate our proposed rule-learner experimentally. After discussing implementation details, and showing a small example for illustration, we then evaluate the performance of our rule-learner on a range of common machine learning datasets comparing it in terms of both accuracy and interpretability to the most popular classification methods. We then confirm the dramatic computational advantages of column-screening and row-sampling on data-sets with a large number of rows and columns.

6.1 Implementation Notes

As discussed in Sect. 3.1, continuous features are approached using indicator functions on thresholds in both directions of comparison; in particular we use 10 quantile-based thresholds per continuous feature dimension. To solve the LP (10), we use IBM CPLEX version 12.4 on a single processor of a 2.33 GHz Intel Xeon-based Linux machine. We find the optimal binary solution via branch and bound. For most of the examples here, the LP itself produces integral solutions. We set the regularization parameter $\lambda = 1/1000$ and do not optimize it in this work. In addition to our single-rule learner, we also consider rule-set learners that we have described in [35]. The set covering approach finds incorrectly classified examples after learning the previous rule, and learns the next rule on these examples only. The

Table 2 Tenfold cross-validation test error on various data sets

	1RULE	RUSC	RUB	DLIST	C5.0	CART	TRBAG	KNN	DISCR	SVM
ILPD	0.2985	0.2985	0.2796	0.3654	0.3053	0.3362	0.2950	0.3019	0.3636	0.3002
IONOS	0.0741	0.0712	0.0798	0.1994	0.0741	0.0997	0.0655	0.1368	0.1425	0.0541
LIVER	0.4609	0.4029	0.3942	0.4522	0.3652	0.3768	0.3101	0.3101	0.3768	0.3217
PARKIN	0.1744	0.1538	0.1590	0.2513	0.1641	0.1282	0.0821	0.1641	0.1641	0.1436
PIMA	0.2617	0.2539	0.2526	0.3138	0.2487	0.2891	0.2305	0.2969	0.2370	0.2344
SONAR	0.3702	0.3137	0.3413	0.3846	0.2500	0.2837	0.1490	0.2260	0.2452	0.1442
TRANS	0.2406	0.2406	0.2420	0.3543	0.2166	0.2701	0.2540	0.2286	0.3369	0.2353
WDBC	0.0703	0.0562	0.0562	0.0967	0.0650	0.0808	0.0422	0.0685	0.0404	0.0228

boosting approach creates a classifier that is a linear combination of our single-rule learners, by emphasizing samples that were incorrectly classified in the previous round. We do not attempt to optimize the number of rounds of set-covering or boosting, but leave it at $T = 5$ having interpretability in mind.

6.2 *Illustrative Example*

We illustrate the types of sparse interpretable rules that are obtained using the proposed rule learner on the Iris data set. We consider the binary problem of classifying iris versicolor from the other two species, setosa and virginica. Of the four features, sepal length, sepal width, petal length, and petal width, the rule that is learned involves only two features and three Boolean expressions:

- petal length ≤ 5.350 cm; AND
- petal width ≤ 1.700 cm; AND
- petal width > 0.875 cm.

6.3 *Classification Performance Comparisons*

As an empirical study, we consider several interpretable classifiers: the proposed Boolean compressed sensing-based single rule learner (1Rule), the set covering approach to extend the proposed rule learner (RuSC), the boosting approach to extend the proposed rule learner (RuB), the decision lists algorithm in SPSS (DList), the C5.0 Release 2.06 algorithm with rule set option in SPSS (C5.0), and the classification and regression trees algorithm in Matlab's classregtree function (CART).[6]

[6]We use IBM SPSS Modeler 14.1 and Matlab R2009a with default settings.

Table 3 Tenfold average number of conjunctive clauses in rule set

	1RULE	RUSC	RUB	DLIST	C5.0
ILPD	1.0	1.2	5.0	3.7	11.7
IONOS	1.0	4.1	5.0	3.7	8.4
LIVER	1.0	3.5	5.0	1.1	15.3
PARKIN	1.0	3.1	5.0	1.2	7.3
PIMA	1.0	2.3	5.0	5.0	12.0
SONAR	1.0	3.9	5.0	1.0	10.4
TRANS	1.0	1.2	5.0	2.3	4.3
WDBC	1.0	4.1	5.0	3.2	7.4

We also consider several classifiers that are not interpretable: the random forests classifier in Matlab's TreeBagger class (TrBag), the k-nearest neighbor algorithm in SPSS (kNN), discriminant analysis of the Matlab function classify, and SVMs with radial basis function kernel in SPSS (SVM).

The data sets to which we apply these classification algorithms come from the UCI repository [22]. They are all binary classification data sets with real-valued features. (We have not considered data sets with categorical-valued features in this study to allow comparisons to a broader set of classifiers; in fact, classification of categorical-valued features is a setting in which rule-based approaches excel.) The specific data sets are: Indian liver patient dataset (ILPD), Ionosphere (Ionos), BUPA liver disorders (Liver), Parkinsons (Parkin), Pima Indian diabetes (Pima), connectionist bench sonar (Sonar), blood transfusion service center (Trans), and breast cancer Wisconsin diagnostic (WDBC).

Table 2 gives tenfold cross-validation test errors for the various classifiers. Table 3 gives the average number of rules across the tenfolds needed by the different rule-based classifiers to achieve those error rates.

It can be noted that our rule sets have better accuracy than decision lists on all data sets and our single rule has better accuracy than decision lists in all but one instance. On about half of the data sets, our set covering rule set has fewer rules than decision lists. Taking the number of rules as an indication of interpretability, we see that our set covering rule set has about the same level of interpretability as decision lists but with better classification accuracy. (We did not optimize the number of rules in boosting.) Even our single rule, which is very interpretable, typically has better accuracy than decision lists with more rules.

Compared to the C5.0 rule set, our proposed rule sets are much more interpretable because they have many fewer rules on average across the data sets considered. The accuracy of C5.0 and our rule sets is on par, as each approach has better accuracy on half of the data sets. The best performing algorithms in terms of accuracy are SVMs and random forests, but we see generally quite competitive accuracy with the advantage of interpretability by the proposed approach. On the ILPD data set, our boosting approach has the best accuracy among all ten classifiers considered.

6.4 Examples of Learned Rules

We now illustrate small rules learned with our approach on the 'WDBC' Breast cancer classification dataset. First we use the LP relaxation in (10) with $\lambda = 2.0$ to find a compact AND-rule. The resulting rule has 5 active clauses, and the resulting train error-rate is 0.065. The clauses are as follows:

- mean texture > 14.13 and
- mean concave points > 0.05 and
- standard error radius > 0.19 and
- standard error area > 13.48 and
- worst area > 475.18.

Next we consider an M-of-N rule for the same data-set, and use the same number of clauses. We note that the LP relaxation in fact produces fractional solutions for this dataset, but only a small proportion of variables is fractional, and the problem can be solved reasonably efficiently using IBM CPLEX in about 8 s. The resulting error-rate is 0.028, an improvement over the AND-rule (N-of-N rule) with the same number of clauses. The rule finds 5 clauses, of which at least 3 need to be active for the classifier to return a positive label:

- mean texture > 16.58
- worst perimeter > 120.26
- worst area > 724.48
- worst smoothness > 0.12
- worst concave points > 0.18.

6.5 Screening Results

In this section, we examine the empirical performance of the screening tests on several data sets from the UCI Machine Learning Repository [22] which have many continuous-valued features: ionosphere ($m = 351$), banknote authentication ($m = 1372$), MAGIC gamma telescope ($m = 19{,}020$), and gas sensor array drift ($m = 13{,}910$). The first three are naturally binary classification problems, whereas the fourth is originally a six class problem that we have converted into a binary problem. The classification problem in ionosphere is to classify whether there is structure in the ionosphere based on radar data from Goose Bay, Labrador, in banknote is to classify genuine and forged banknotes from statistics of wavelet-transformed images, in MAGIC is a signal detection task from measurements of the Cherenkov gamma telescope, and in gas is to classify pure gaseous substances using measurements from different chemical sensors. In gas, we map the class labels ammonia, acetaldehyde and acetone to the binary class 0 and ethylene, ethanol and toluene to the binary class 1.

Table 4 Screening results for all pairs column comparison and enhanced primal and dual heuristics

Data set	Features	Thresholds	Columns	Columns screened by simple tests	Columns screened by duality test	Total columns screened	Fraction columns screened
Ionosphere	33	10	642	596	638	638	0.994
		20	1282	1210	1271	1271	0.991
		50	3202	3102	2984	3133	0.978
		100	6402	6269	5968	6310	0.986
Banknote	4	10	80	40	71	71	0.888
		20	160	92	142	142	0.888
		50	400	259	350	355	0.888
		100	800	548	700	712	0.890
MAGIC	10	10	200	188	183	188	0.940
		20	400	375	369	377	0.943
		50	1000	945	916	945	0.945
		100	2000	1892	1799	1892	0.946
Gas	128	10	2560	1875	2242	2256	0.881
		20	5120	3943	4684	4758	0.929
		50	12,800	10,235	11,378	11,824	0.924
		100	25,600	20,935	22,814	23,893	0.933

Table 4 gives the results with the enhanced version of the screening tests described in Sect. 5 that compares all pairs of columns in the pairwise test. The tables shows results for the four data sets with four different numbers of thresholds per feature dimension. We construct the $a_j(\mathbf{x})$ functions by quantile-based thresholds, and consider both directions of Boolean functions, e.g., ‘weight $\leq$ 100’ as well as ‘weight $>$ 100.’ The results show the number of columns screened by the simple tests alone, the number of columns screened by the duality-based test alone, and their union in total columns screened. The tests may also be run sequentially, but for brevity we do not discuss this here.

The first thing to note in the tables is that our screening tests dramatically reduce the number of columns in the LP, ranging from removing 90% to over 99% of columns of the matrix $\mathbf{A}$. The fraction of columns screened is fairly stable across the number of thresholds within a data set, but tends to slightly improve with more thresholds. The simple tests and duality-based tests tend to have a good deal of overlap, but there is no pattern with one being a superset of the other.

The implications for running time are presented in Table 5, where we focus on the largest data set, gas. The first column shows the full LP without any screening. We compare that to the total time for screening and solving the reduced LP for the basic and enhanced screening tests. We can see that screening dramatically reduces the total solution time for the LP. Enhanced screening, while requiring more computation, does compensate the LP time and significantly reduces the total running time. With 100 thresholds we solve a very large binary integer problem with 25,600 variables to optimality in under 15 s.

Table 5 Gas data set running times in seconds for screening, solving the LP, and the total of the two: (a) basic tests, and (b) enhanced tests

Thr.	Full LP	(a) Scr.	(a) LP	(a) Tot.	(b) Scr.	(b) LP	(b) Tot.
10	18.58	0.34	2.47	2.81	0.74	1.38	2.12
20	39.52	0.73	3.96	4.69	1.53	1.29	2.82
50	103.46	2.01	12.12	14.13	4.03	3.56	7.59
100	215.57	4.28	24.30	28.58	8.86	5.90	14.76

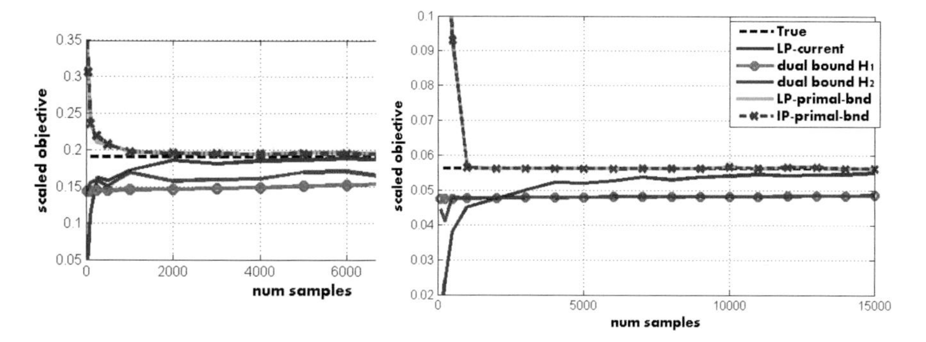

Fig. 2 Illustration of upper and lower bounds on the rule-learning LP and IP objective values for the UCI Adult (**a**) and Census Income (**b**) classification datasets. We obtain tight bounds using only a small fraction of the data.

6.6 Row Sampling Results

In this section, we apply our row-sampling bounds from Sect. 5.3 to two large-scale binary classification tasks from the UCI Machine Learning Repository [22] with a large number of rows. We consider the "Adult" and the "Census Income" datasets, which come with 32,560 and 199,522 training samples respectively. After converting categorical and continuous features to binary (using 10 thresholds) the 101 features in "Adult" dataset and the 354 features in the "Census income" dataset produce 310 and 812 columns in the corresponding $\mathbf{A}$-matrix representations.

The results for both datasets are illustrated in Fig. 2. We plot our various bounds as a function of m: we show the objective value of the full LP (constant dashed line), and of the small LP, the upper bounds on both the LP and IP solutions for the full dataset, and the two dual bounds. We can see that the objective value of the small LP and both the LP and IP upper bounds quickly approach the objective value of the full LP (after about 2000 samples out of 32,560 total for the Adult dataset, and after 5000 samples out of 199,522 total for the bigger "Census income"). The dual bounds improve with time, albeit slower than the upper bounds. The second dual extension approach provides a much tighter lower bound for the "Adult" dataset in plot (a), but only a very modest gain for "Census Income" in plot (b). Remarkably, for both UCI examples the LP and IP solutions for the small LP are either the same or very close, allowing quick integral solution via branch and bound. The same holds for the LP and IP upper bounds.

7 Conclusion

In this chapter, we have developed a new approach for learning decision rules based on compressed sensing ideas. The approach leads to a powerful rule-based classification system whose outputs are easy for human users to trust and draw insight from. In contrast to typical rule learners, the proposed approach is not heuristic. We prove theoretical results showing that exact rule recovery is possible through a convex relaxation of the combinatorial optimization problem under certain conditions. We also extend the framework to a more general classification problem of learning checklists and scorecards, using M-of-N rule formulation. We extend the LP formulation, and the theoretical results using ideas from Threshold Group Testing.

For large scale classification problems, we have developed novel screening tests and row sampling approaches. One class of screening tests is based on counting false alarm and missed detection errors whereas the other class is based on duality theory. In contrast to Lasso screening, which makes use of strong duality, the proposed tests consider the integer nature of the Boolean compressed sensing problem to check if the dual value is less than or equal to the optimal integer value. We developed LP duality-based techniques to guarantee a near-optimal solution after training the classifier only on a small subset of the available samples in both batch and online settings.

In our empirical evaluation we have shown that the proposed algorithm is practical and leads to a classifier that has a better trade-off of accuracy with interpretability than existing classification approaches. It produces better accuracy than existing interpretable classifiers, and much better interpretability than the powerful black-box classifiers such as SVMs and random forests while typically paying only minor cost in accuracy. Furthermore, our experimental evaluation confirms the significant gains in computational complexity of the proposed screening and row-sampling approaches.

Appendix 1: Dual Linear Program

We now derive the dual LP, which we use in Sect. 5. We start off by giving a reformulation of the LP in (10), i.e., we consider an LP with the same set of optimal solutions as the one in (10). First note that the upper bounds of 1 on the variables ξ_i are redundant. Let $(\bar{\mathbf{w}}, \bar{\boldsymbol{\xi}})$ be a feasible solution of (10) without the upper bound constraints such that $\bar{\xi}_i > 1$ for some $i \in \mathscr{P}$. Reducing $\bar{\xi}_i$ to 1 yields a feasible solution (as $\mathbf{a}_i\bar{\mathbf{w}} + \bar{\xi}_i \geq 1$—the only inequality ξ_i participates in besides the bound constraints—is still satisfied). The new feasible solution has lower objective function value than before, as ξ_i has a positive coefficient in the objective function (which is to be minimized). One can similarly argue that in every optimal solution of (10) without the upper bound constraints, we have $w_j \leq 1$ (for $j = 1, \ldots, n$). Finally, observe that we can substitute ξ_i for $i \in \mathscr{Z}$ in the objective function by $\mathbf{a}_i\mathbf{w}$ because of the constraints $\mathbf{a}_i\mathbf{w} = \xi_i$ for $i \in \mathscr{Z}$. We thus get the following LP equivalent to (10):

$$\begin{aligned} \min \quad & \sum_{j=1}^{n} \left(\lambda + \|\mathbf{a}^j_{\mathscr{Z}}\|_1\right) w_j + \sum_{i=1}^{p} \xi_i \\ \text{s.t.} \quad & 0 \leq w_j,\ j = 1, \ldots, n \\ & 0 \leq \xi_i,\ i = 1, \ldots, p \\ & \mathbf{A}_{\mathscr{P}}\mathbf{w} + \boldsymbol{\xi}_{\mathscr{P}} \geq \mathbf{1}. \end{aligned} \tag{19}$$

The optimal solutions and optimal objective values are the same as in (10).

Writing $\mathbf{A}_{\mathscr{P}}\mathbf{w} + \boldsymbol{\xi}_{\mathscr{P}}$ as $\mathbf{A}_{\mathscr{P}}\mathbf{w} + \mathbf{I}\boldsymbol{\xi}_{\mathscr{P}}$, where $\mathbf{I}$ is the $p \times p$ identity matrix, $||\mathbf{a}^j_{\mathscr{Z}}||_1$ as $\mathbf{1}^T\mathbf{a}^j_{\mathscr{Z}}$, and letting $\boldsymbol{\mu}$ be a row vector of p dual variables, one can see that the dual is:

$$\begin{aligned} \max \quad & \sum_{i=1}^{p} \mu_i \\ \text{s.t.} \quad & 0 \leq \mu_i \leq 1,\ i = 1, \ldots, p \\ & \boldsymbol{\mu}^T\mathbf{A}_{\mathscr{P}} \leq \lambda\mathbf{1}_n + \mathbf{1}^T\mathbf{A}_{\mathscr{Z}}. \end{aligned} \tag{20}$$

Suppose $\bar{\boldsymbol{\mu}}$ is a feasible solution to (20). Then clearly $\sum_{i=1}^{p} \bar{\mu}_i$ yields a lower bound on the optimal solution value of (19).

Appendix 2: Derivation of Screening Tests

Let $\mathscr{S}(j)$ stand for the support of $\mathbf{a}^j_{\mathscr{P}}$. Furthermore, let $\mathscr{N}(j)$ stand for the support of $\mathbf{1} - \mathbf{a}^j_{\mathscr{P}}$, i.e, it is the set of indices from $\mathscr{P}$ such that the corresponding components of $\mathbf{a}^j_{\mathscr{P}}$ are zero.

Now consider the situation where we fix w_1 (say) to 1. Let $\mathbf{A}'$ stand for the submatrix of $\mathbf{A}$ consisting of the last $n-1$ columns. Let $\mathbf{w}'$ stand for the vector of variables $w_2, \ldots, w_n$. Then the constraints $\mathbf{A}_{\mathscr{P}}\mathbf{w} + \boldsymbol{\xi}_{\mathscr{P}} \geq \mathbf{1}$ in (19) become $\mathbf{A}'_{\mathscr{P}}\mathbf{w}' + \boldsymbol{\xi}_{\mathscr{P}} \geq \mathbf{1} - \mathbf{a}^1_{\mathscr{P}}$. Therefore, for all $i \in \mathscr{S}(1)$, the corresponding constraint is now $(\mathbf{A}'_{\mathscr{P}})_i\mathbf{w}' + \xi_i \geq 0$ which is a redundant constraint as $\mathbf{A}'_{\mathscr{P}} \geq 0$ and $\mathbf{w}', \xi_i \geq 0$.

The only remaining non-redundant constraints correspond to the indices in $\mathcal{N}(1)$. Then the value of (19) with w_1 set to 1 becomes

$$
\begin{aligned}
(\lambda + \|\mathbf{a}_{\mathcal{Z}}^1\|_1) + \min \quad & \sum_{j=2}^{n} \left(\lambda + \|\mathbf{a}_{\mathcal{Z}}^j\|_1\right) w_j + \sum_{i \in \mathcal{N}(1)} \xi_i \\
\text{s.t.} \quad & 0 \le w_j,\ j = 2, \ldots, n \\
& 0 \le \xi_i,\ i \in \mathcal{N}(1) \\
& \mathbf{A}'_{\mathcal{N}(1)} \mathbf{w}' + \boldsymbol{\xi}_{\mathcal{N}(1)} \ge \mathbf{1}.
\end{aligned}
\tag{21}
$$

This LP clearly has the same form as the LP in (19). Furthermore, given any feasible solution $\bar{\boldsymbol{\mu}}$ of (20), $\bar{\boldsymbol{\mu}}_{\mathcal{N}(1)}$ defines a feasible dual solution of (21) as

$$
\begin{aligned}
& \bar{\boldsymbol{\mu}}^T \mathbf{A}_{\mathcal{P}} \le \lambda \mathbf{1}_n + \mathbf{1}^T \mathbf{A}_{\mathcal{Z}} \\
\Rightarrow\ & \bar{\boldsymbol{\mu}}_{\mathcal{S}(1)}^T \mathbf{A}'_{\mathcal{S}(1)} + \bar{\boldsymbol{\mu}}_{\mathcal{N}(1)}^T \mathbf{A}'_{\mathcal{N}(1)} \le \lambda \mathbf{1}_{n-1} + \mathbf{1}^T \mathbf{A}'_{\mathcal{Z}} \\
\Rightarrow\ & \bar{\boldsymbol{\mu}}_{\mathcal{N}(1)}^T \mathbf{A}'_{\mathcal{N}(1)} \le \lambda \mathbf{1}_{n-1} + \mathbf{1}^T \mathbf{A}'_{\mathcal{Z}}.
\end{aligned}
$$

Therefore $\sum_{i \in \mathcal{N}(n)} \bar{\mu}_i$ is a lower bound on the optimal solution value of the LP in (21) and therefore

$$
\lambda + ||\mathbf{a}_{\mathcal{Z}}^1||_1 + \sum_{i \in \mathcal{N}(1)} \bar{\mu}_i \tag{22}
$$

is a lower bound on the optimal solution value of (19) with w_1 set to 1. In particular, if $(\bar{\mathbf{w}}, \bar{\boldsymbol{\xi}})$ is a feasible *integral* solution to (19) with objective function value $\lambda(\sum_{i=1}^n \bar{w}_i) + \sum_{i=1}^p \bar{\xi}_i$, and if (22) is greater than this value, than no optimal integral solution of (19) can have $w_1 = 1$. Therefore $w_1 = 0$ in any optimal solution, and we can simply drop the column corresponding to w_1 from the LP.

In order to use the screening results in this section we need to obtain a feasible primal and a feasible dual solution. Some useful heuristics to obtain such a pair are described in [12].

Appendix 3: Extending the Dual Solution for Row-Sampling

Suppose that $\hat{\boldsymbol{\mu}}^p$ is the optimal dual solution to the small LP in Sect. 5.3. Note that the number of variables in the dual for the large LP increases from p to $\bar{p}$ and the bound on the second constraint grows from $\lambda \mathbf{1}_n + \mathbf{1}^T \mathbf{A}_{\mathcal{Z}}$ to $\lambda \mathbf{1}_n + \mathbf{1}^T \bar{\mathbf{A}}_{\mathcal{Z}}$.

We use a greedy heuristic to extend $\hat{\boldsymbol{\mu}}^p$ to a feasible dual solution $\bar{\boldsymbol{\mu}}_{\bar{p}}$ of the large LP. We set $\bar{\mu}_j = \hat{\mu}_j$ for $j = 1, .., p$. We extend the remaining entries $\bar{\mu}_j$ for $j = (p+1), .., \bar{p}$ by setting a subset of its entries to 1 while satisfying the dual feasibility constraint. In other words the extension of $\bar{\boldsymbol{\mu}}$ corresponds to a subset $\mathscr{R}$ of the row indices $\{p+1,\ldots,\bar{p}\}$ of $\bar{\mathbf{A}}_{\mathscr{P}}$ such that $\hat{\boldsymbol{\mu}}_p^T \mathbf{A}_{\mathscr{P}} + \sum_{i\in\mathscr{R}} (\bar{\mathbf{A}}_{\mathscr{P}})_i \leq \mathbf{1}^T \bar{\mathbf{A}}_{\mathscr{Z}}$. Having $\bar{\boldsymbol{\mu}}^T \mathbf{A}_{\mathscr{P}} \leq \mathbf{1}^T \mathbf{A}_{\mathscr{Z}}$ with $\bar{\boldsymbol{\mu}}$ extended by a binary vector implies that $\bar{\boldsymbol{\mu}}$ is feasible for (20). We initialize $\mathscr{R}$ to $\emptyset$ and then simply go through the unseen rows of $\bar{\mathbf{A}}_{\mathscr{P}}$ in some fixed order (increasing from $p+1$ to $\bar{p}$), and for a row k, if

$$\hat{\boldsymbol{\mu}}_p^T \mathbf{A}_{\mathscr{P}} + \sum_{i\in\mathscr{R}} (\bar{\mathbf{A}}_{\mathscr{P}})_i + (\bar{\mathbf{A}}_{\mathscr{P}})_k \leq \mathbf{1}^T \bar{\mathbf{A}}_{\mathscr{Z}},$$

we set $\mathscr{R}$ to $\mathscr{R} \cup \{k\}$. The heuristic (we call it H1) needs only a single pass through the matrix $\bar{\mathbf{A}}_{\mathscr{P}}$, and is thus very fast.

This heuristic, however, does not use the optimal solution $\hat{\mathbf{w}}^m$ in any way. Suppose $\hat{\mathbf{w}}^m$ were an optimal solution of the large LP. Then complementary slackness would imply that if $(\bar{\mathbf{A}}_{\mathscr{P}})_i \hat{\mathbf{w}}^m > 1$, then in any optimal dual solution $\boldsymbol{\mu}$, $\mu_i = 0$. Thus, assuming $\hat{\mathbf{w}}^m$ is close to an optimal solution for the large LP, we modify heuristic H1 to obtain heuristic H2, by simply setting $\bar{\mu}_i = 0$ whenever $(\bar{\mathbf{A}}_{\mathscr{P}})_i \hat{\mathbf{w}}^m > 1$, while keeping the remaining steps unchanged.

Acknowledgements The authors thank Vijay S. Iyengar, Benjamin Letham, Cynthia Rudin, Viswanath Nagarajan, Karthikeyan Natesan Ramamurthy, Mikhail Malyutov and Venkatesh Saligrama for valuable discussions.

References

1. Adams, S.T., Leveson, S.H.: Clinical prediction rules. Br. Med. J. **344**, d8312 (2012)
2. Atia, G.K., Saligrama, V.: Boolean compressed sensing and noisy group testing. IEEE Trans. Inf. Theory **58**(3), 1880–1901 (2012)
3. Bertsimas, D., Chang, A., Rudin, C.: An integer optimization approach to associative classification. In: Advances in Neural Information Processing Systems 25, pp. 269–277 (2012)
4. Blum, A., Kalai, A., Langford, J.: Beating the hold-out: bounds for k-fold and progressive cross-validation. In: Proceedings of the Conference on Computational Learning Theory, Santa Cruz, CA, pp. 203–208 (1999)
5. Boros, E., Hammer, P.L., Ibaraki, T., Kogan, A., Mayoraz, E., Muchnik, I.: An implementation of logical analysis of data. IEEE Trans. Knowl. Data Eng. **12**(2), 292–306 (2000)
6. Candès, E.J., Wakin, M.B.: An introduction to compressive sampling. IEEE Signal Process. Mag. **25**(2), 21–30 (2008)
7. Chen, H.B., Fu, H.L.: Nonadaptive algorithms for threshold group testing. Discret. Appl. Math. **157**, 1581–1585 (2009)
8. Cheraghchi, M., Hormati, A., Karbasi, A., Vetterli, M.: Compressed sensing with probabilistic measurements: a group testing solution. In: Proceedings of the Annual Allerton Conference on Communication Control and Computing, Allerton, IL, pp. 30–35 (2009)
9. Clark, P., Niblett, T.: The CN2 induction algorithm. Mach. Learn. **3**(4), 261–283 (1989)
10. Cohen, W.W.: Fast effective rule induction. In: Proceedings of the International Conference on Machine Learning, Tahoe City, CA, pp. 115–123 (1995)
11. Dai, L., Pelckmans, K.: An ellipsoid based, two-stage screening test for BPDN. In: Proceedings of the European Signal Processing Conference, Bucharest, Romania, pp. 654–658 (2012)

12. Dash, S., Malioutov, D.M., Varshney, K.R.: Screening for learning classification rules via Boolean compressed sensing. In: Proceedings of the IEEE International Conference on Acoustics, Speech, and Signal Processing, Florence, Italy, pp. 3360–3364 (2014)
13. Dash, S., Malioutov, D.M., Varshney, K.R.: Learning interpretable classification rules using sequential row sampling. In: Proceedings of the IEEE International Conference on Acoustics, Speech, and Signal Processing, Brisbane, Australia (2015)
14. Dembczyński, K., Kotłowski, W., Słowiński, R.: ENDER: a statistical framework for boosting decision rules. Data Min. Knowl. Disc. **21**(1), 52–90 (2010)
15. Donoho, D.L., Elad, M.: Optimally sparse representation in general (nonorthogonal) dictionaries via l1 minimization. Proc. Natl. Acad. Sci. **100**(5), 2197–2202 (2003)
16. Du, D.Z., Hwang, F.K.: Pooling Designs and Nonadaptive Group Testing: Important Tools for DNA Sequencing. World Scientific, Singapore (2006)
17. Dyachkov, A.G., Rykov, V.V.: A survey of superimposed code theory. Prob. Control. Inf. **12**(4), 229–242 (1983)
18. Dyachkov, A.G., Vilenkin, P.A., Macula, A.J., Torney, D.C.: Families of finite sets in which no intersection of l sets is covered by the union of s others. J. Combin. Theory **99**, 195–218 (2002)
19. Eckstein, J., Goldberg, N.: An improved branch-and-bound method for maximum monomial agreement. INFORMS J. Comput. **24**(2), 328–341 (2012)
20. El Ghaoui, L., Viallon, V., Rabbani, T.: Safe feature elimination in sparse supervised learning. Pac. J. Optim. **8**(4), 667–698 (2012)
21. Emad, A., Milenkovic, O.: Semiquantitative group testing. IEEE Trans. Inf. Theory **60**(8), 4614–4636 (2014)
22. Frank, A., Asuncion, A.: UCI machine learning repository. http://archive.ics.uci.edu/ml (2010)
23. Friedman, J.H., Popescu, B.E.: Predictive learning via rule ensembles. Ann. Appl. Stat. **2**(3), 916–954 (2008)
24. Fry, C.: Closing the gap between analytics and action. INFORMS Analytics Mag. **4**(6), 4–5 (2011)
25. Gage, B.F., Waterman, A.D., Shannon, W., Boechler, M., Rich, M.W., Radford, M.J.: Validation of clinical classification schemes for predicting stroke. J. Am. Med. Assoc. **258**(22), 2864–2870 (2001)
26. Gawande, A.: The Checklist Manifesto: How To Get Things Right. Metropolitan Books, New York (2009)
27. Gilbert, A.C., Iwen, M.A., Strauss, M.J.: Group testing and sparse signal recovery. In: Conference Record - Asilomar Conference on Signals, Systems and Computers, Pacific Grove, CA, pp. 1059–1063 (2008)
28. Jawanpuria, P., Nath, J.S., Ramakrishnan, G.: Efficient rule ensemble learning using hierarchical kernels. In: Proceedings of the International Conference on Machine Learning, Bellevue, WA, pp. 161–168 (2011)
29. John, G.H., Langley, P.: Static versus dynamic sampling for data mining. In: Proceedings of the ACM SIGKDD International Conference on Knowledge Discovery and Data Mining, Portland, OR, pp. 367–370 (1996)
30. Kautz, W., Singleton, R.: Nonrandom binary superimposed codes. IEEE Trans. Inf. Theory **10**(4), 363–377 (1964)
31. Letham, B., Rudin, C., McCormick, T.H., Madigan, D.: Building interpretable classifiers with rules using Bayesian analysis. Tech. Rep. 609, Department of Statistics, University of Washington (2012)
32. Liu, J., Li, M.: Finding cancer biomarkers from mass spectrometry data by decision lists. J. Comput. Biol. **12**(7), 971–979 (2005)
33. Liu, J., Zhao, Z., Wang, J., Ye, J.: Safe screening with variational inequalities and its application to lasso. In: Proceedings of the International Conference on Machine Learning, Beijing, China, pp. 289–297 (2014)

34. Malioutov, D., Malyutov, M.: Boolean compressed sensing: LP relaxation for group testing. In: Proceedings of the IEEE International Conference on Acoustics, Speech and Signal Processing, Kyoto, Japan, pp. 3305–3308 (2012)
35. Malioutov, D.M., Varshney, K.R.: Exact rule learning via Boolean compressed sensing. In: Proceedings of the International Conference on Machine Learning, Atlanta, GA, pp. 765–773 (2013)
36. Malioutov, D.M., Sanghavi, S.R., Willsky, A.S.: Sequential compressed sensing. IEEE J. Spec. Top. Signal Proc. **4**(2), 435–444 (2010)
37. Malyutov, M.: The separating property of random matrices. Math. Notes **23**(1), 84–91 (1978)
38. Malyutov, M.: Search for sparse active inputs: a review. In: Aydinian, H., Cicalese, F., Deppe, C. (eds.) Information Theory, Combinatorics, and Search Theory: In Memory of Rudolf Ahlswede, pp. 609–647. Springer, Berlin/Germany (2013)
39. Marchand, M., Shawe-Taylor, J.: The set covering machine. J. Mach. Learn. Res. **3**, 723–746 (2002)
40. Maron, O., Moore, A.W.: Hoeffding races: accelerating model selection search for classification and function approximation. Adv. Neural Inf. Proces. Syst. **6**, 59–66 (1993)
41. Mazumdar, A.: On almost disjunct matrices for group testing. In: Proceedings of the International Symposium on Algorithms and Computation, Taipei, Taiwan, pp. 649–658 (2012)
42. Provost, F., Jensen, D., Oates, T.: Efficient progressive sampling. In: Proceedings of the ACM SIGKDD International Conference on Knowledge Discovery and Data Mining, San Diego, CA, pp. 23–32 (1999)
43. Quinlan, J.R.: Simplifying decision trees. Int. J. Man Mach. Stud. **27**(3), 221–234 (1987)
44. Rivest, R.L.: Learning decision lists. Mach. Learn. **2**(3), 229–246 (1987)
45. Rückert, U., Kramer, S.: Margin-based first-order rule learning. Mach. Learn. **70**(2–3), 189–206 (2008)
46. Sejdinovic, D., Johnson, O.: Note on noisy group testing: asymptotic bounds and belief propagation reconstruction. In: Proceedings of the Annual Allerton Conference on Communication Control and Computing, Allerton, IL, pp. 998–1003 (2010)
47. Stinson, D.R., Wei, R.: Generalized cover-free families. Discret. Math. **279**, 463–477 (2004)
48. Ustun, B., Rudin, C.: Methods and models for interpretable linear classification. Available at http://arxiv.org/pdf/1405.4047 (2014)
49. Wagstaff, K.L.: Machine learning that matters. In: Proceedings of the International Conference on Machine Learning, Edinburgh, United Kingdom, pp. 529–536 (2012)
50. Wang, F., Rudin, C.: Falling rule lists. Available at http://arxiv.org/pdf/1411.5899 (2014)
51. Wang, J., Zhou, J., Wonka, P., Ye, J.: Lasso screening rules via dual polytope projection. Adv. Neural Inf. Proces. Syst. **26**, 1070–1078 (2013)
52. Wang, Y., Xiang, Z.J., Ramadge, P.J.: Lasso screening with a small regularization parameter. In: Proceedings of the IEEE International Conference on Acoustics, Speech and Signal Processing, Vancouver, Canada, pp. 3342–3346 (2013)
53. Wang, Y., Xiang, Z.J., Ramadge, P.J.: Tradeoffs in improved screening of lasso problems. In: Proceedings of the IEEE International Conference on Acoustics, Speech and Signal Processing, Vancouver, Canada, pp. 3297–3301 (2013)
54. Wang, T., Rudin, C., Doshi, F., Liu, Y., Klampfl, E., MacNeille, P.: Bayesian or's of and's for interpretable classification with application to context aware recommender systems. Available at http://arxiv.org/abs/1504.07614 (2015)
55. Wu, H., Ramadge, P.J.: The 2-codeword screening test for lasso problems. In: Proceedings of the IEEE International Conference on Acoustics, Speech and Signal Processing, Vancouver, Canada, pp. 3307–3311 (2013)
56. Xiang, Z.J., Ramadge, P.J.: Fast lasso screening tests based on correlations. In: Proceedings of the IEEE International Conference on Acoustics, Speech and Signal Processing, Kyoto, Japan, pp. 2137–2140 (2012)
57. Xiang, Z.J., Xu, H., Ramadge, P.J.: Learning Sparse Representations of High Dimensional Data on Large Scale Dictionaries. Advances in Neural Information Processing Systems, vol. 24, pp. 900–908. MIT Press, Cambridge, MA (2011)

Visualizations of Deep Neural Networks in Computer Vision: A Survey

Christin Seifert, Aisha Aamir, Aparna Balagopalan, Dhruv Jain, Abhinav Sharma, Sebastian Grottel, and Stefan Gumhold

Abstract In recent years, Deep Neural Networks (DNNs) have been shown to outperform the state-of-the-art in multiple areas, such as visual object recognition, genomics and speech recognition. Due to the distributed encodings of information, DNNs are hard to understand and interpret. To this end, visualizations have been used to understand how deep architecture work in general, what different layers of the network encode, what the limitations of the trained model was and how to interactively collect user feedback. In this chapter, we provide a survey of visualizations of DNNs in the field of computer vision. We define a classification scheme describing visualization goals and methods as well as the application areas. This survey gives an overview of what can be learned from visualizing DNNs and which visualization methods were used to gain which insights. We found that most papers use Pixel Displays to show neuron activations. However, recently more sophisticated visualizations like interactive node-link diagrams were proposed. The presented overview can serve as a guideline when applying visualizations while designing DNNs.

Network Architecture Type Abbreviations

CDBN	Convolutional Deep Belief Networks
CNN	Convolutional Neural Networks
DBN	Deep Belief Networks
DCNN	Deep Convolution Neural Network
DNNs	Deep Neural Networks
MCDNN	Multicolumn Deep Neural Networks

C. Seifert (✉) • A. Aamir • A. Balagopalan • D. Jain • A. Sharma • S. Grottel • S. Gumhold
Technische Universität Dresden, Dresden, Germany
e-mail: Christin.42.Seifert@gmail.com; aishaaamir7@gmail.com; aparna.balagopalan@gmail.com; dhruvjain.1027@gmail.com; abhinav0301@gmail.com; sebastian.grottel@tu-dresden.de; stefan.gumhold@tu-dresden.de

T. Cerquitelli et al. (eds.), *Transparent Data Mining for Big and Small Data*, Studies in Big Data 32, DOI 10.1007/978-3-319-54024-5_6

Dataset Name Abbreviations

CASIA-HWDB Institute of Automation of Chinese Academy of Sciences-Hand Writing Databases
DTD Describable Textures Dataset
FLIC Frames Labeled In Cinema
FMD Flickr Material Database
GTSRB German Traffic Sign Recognition Benchmark
ISLVRC ImageNet Large Scale Visual Recognition Challenge
LFW Labeled Faces in the Wild
LSP Leeds Sports Pose
MNIST Mixed National Institute of Standards and Technology
VOC Visual Object Classes
WAF We Are Family
YTF YouTube Faces

Other Abbreviations

CVPR Computer Vision and Pattern Recognition
ICCV International Conference on Computer Vision
IEEE Institute of Electrical and Electronics Engineers
NIPS Neural Information Processing Systems
t-SNE Stochastic Neighbor Embedding

1 Introduction

Artificial Neural Networks for learning mathematical functions have been introduced in 1943 [48]. Despite being theoretically able to approximate any function [7], their popularity decreased in the 1970s because their computationally expensive training was not feasible with available computing resources [49]. With the increase in computing power in recent years, neural networks again became subject of research as Deep Neural Networks (DNNs). DNNs, artificial neural networks with multiple layers combining supervised and unsupervised training, have since been shown to outperform the state-of-the-art in multiple areas, such as visual object recognition, genomics and speech recognition [36]. Despite their empirically superior performance, DNN models have one disadvantage: their trained models are not easily understandable, because information is encoded in a distributed manner.

However, understanding and trust have been identified as desirable property of data mining models [65]. In most scenarios, experts can assess model performance on data sets, including gold standard data sets, but have little insights on how and why a specific model works [81]. The missing understandability is one of the reasons why less powerful, but easy to communicate classification models such as decision trees are in some applications preferred to very powerful classification models, like Support Vector Machines and Artificial Neural Networks [33]. Visualization has been shown to support understandability for various data mining models, e.g. for Naive Bayes [1] and Decision Forests [66].

In this chapter, we review literature on visualization of DNNs in the computer vision domain. Although DNNs have many application areas, including automatic translation and text generation, computer vision tasks are the earliest applications [35]. Computer vision applications also provide the most visualization possibilities due to their easy-to-visualize input data, i.e., images. In the review, we identify questions authors ask about neural networks that should be answered by a visualization (*visualization goal*) and which *visualization methods* they apply therefore. We also characterize the application domain by the computer vision *task* the network is trained for, the *type* of network architecture and the *data sets* used for training and visualization. Note that we only consider visualizations which are automatically generated. We do not cover manually generated illustrations (like the network architecture illustration in [35]). Concretely, our research questions are:

RQ-1 Which insights can be gained about DNN models by means of visualization?
RQ-2 Which visualization methods are appropriate for which kind of insights?

To collect the literature we pursued the following steps: since deep architectures became prominent only a few years ago, we restricted our search starting from the year 2010. We searched the main conferences, journals and workshops in the area of computer vision, machine learning and visualization, such as: IEEE International Conference on Computer Vision (ICCV), IEEE Conferences on Computer Vision and Pattern Recognition (CVPR), IEEE Visualization Conference (VIS), Advances in Neural Information Processing Systems (NIPS). Additionally, we used keyword-based search in academic search engines, using the following phrases (and combinations): "deep neural networks", "dnn", "visualization", "visual analysis", "visual representation", "feature visualization".

This chapter is organized as follows: the next section introduces the classification scheme and describes the categories we applied to the collected papers. Section 3 reviews the literature according to the introduced categories. We discuss the findings with respect to the introduced research questions in Sect. 4, and conclude the work in Sect. 5.

2 Classification Scheme

In this chapter we present the classification scheme used to structure the literature: we first introduce a general view, and then provide detailed descriptions of the categories and their values. An overview of the classification scheme is shown in Fig. 1.

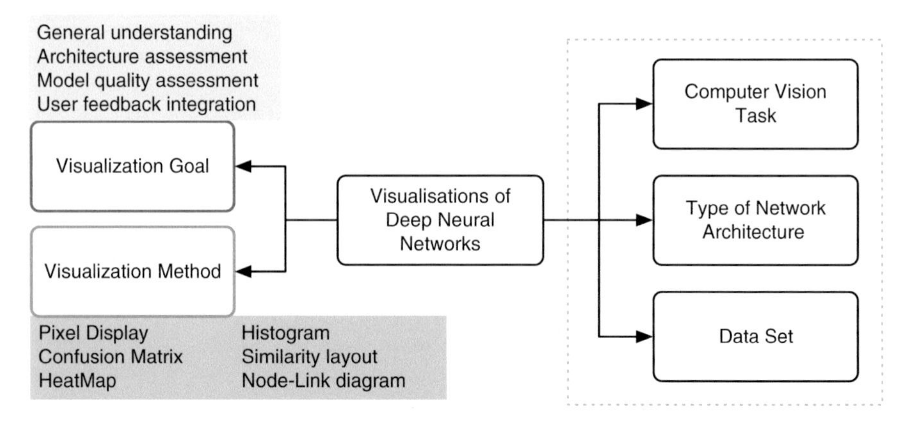

Fig. 1 Classification Scheme for Visualizations of Deep Neural Networks. The *dotted border* subsumes the categories characterizing the application area

First, we need to identify the purpose the visualization was developed for. We call this category **visualization goal**. Possible values are for instance *general understanding* and *model quality assessment*. Then, we identified the **visualization methods** used to achieve the above mentioned goals. Such methods can potentially cover the whole visualization space [51], but literature review shows that only a very small subset has been used so far in the context of DNNs, including *heat maps* and visualizations of *confusion matrices*. Additionally, we introduced three categories to describe the application domain. These categories are the *computer vision task*, the *architecture type* of the network and the *data sets* the neural network was trained on, which is also used for the visualization.

Note, that the categorization is not distinct. This means that one paper can be assigned multiple values in one category. For instance, a paper can use multiple visualization methods (CNNVis uses a combination of node-link diagrams, matrix displays and heatmaps [44]) on multiple data sets.

Related to the proposed classification scheme is the taxonomy of Grün et al. for visualizing learned features in convolutional neural networks [25]. The authors categorize the visualization methods into *input modification*, *de-convolutional*, and *input reconstruction* methods. In input modification methods, the output of the network and intermediate layers is measured while the input is modified. De-Convolutional methods adapt a reverse strategy to calculate the influence of a neuron's activation from lower layers. This strategy demonstrates which pixels are responsible for the activation of neurons in each layer of the network. Input reconstruction methods try to assess the importance of features by reconstructing input images. These input images can either be real or artificial images, that either maximize or lead to an output invariance of a unit of interest. This categorization is restricted to feature visualizations and therefore is narrower as the proposed scheme. For instance, it does not cover the general application domain, and is restricted to specific types of visualizations, because it categorizes the calculation methods used for Pixel Displays and heatmaps.

2.1 Visualization Goals

This category describes the various goals of the authors visualizing DNNs. We identified the following four main goals:

- *General Understanding:* This category encompasses questions about general behavior of the neural network, either during training, on the evaluation data set or on unseen images. Authors want to find out what different network layers are learning or have learned, on a rather general level.
- *Architecture Assessment:* Work in this category tries to identify how the network architecture influences performance in detail. Compared to the first category the analyses are on a more fine-grained level, e.g. assessing which layers of the architecture represent which features (e.g., color, texture), and which feature combinations are the basis for the final decision.
- *Model Quality Assessment:* In this category authors have focused their research goal in determining how the number of layers and role played by each layer can affect the visualization process.
- *User Feedback Integration:* This category comprises work in which visualization is the means to integrate user feedback into the machine learning model. Examples for such feedback integration are user-based selection of training data [58] or the interactive refinement of hypotheses [21].

2.2 Visualization Methods

Only a few visualization methods [51] have been applied to DNNs. We briefly describe them in the following:

- *Histogram:* A histogram is a very basic visualization showing the distribution of univariate data as a bar chart.
- *Pixel Displays:* The basic idea is that each pixel represents a data point. In the context of DNN, the (color) value for each pixel is based on network activation, reconstructions, or similar and yield 2-dimensional rectangular images. In most cases the pixels next to each other in the display space are also next to each other in the semantic space (e.g., nearby pixels of the original image). This nearness criterion is defined on the difference from Dense Pixel Displays [32]. We further distinguish whether the displayed values originate from a single image, from a set of images (i.e., a batch), or only from a part of the image.
- *Heat Maps:* Heat maps are a special case of Pixel Displays, where the value for each pixel represents an accumulated quantity of some kind and is encoded using a specific coloring scheme [72]. Heat maps are often transparently overlaid over the original data.
- *Similarity Layout:* In similarity-based layouts the relative positions of data objects in the low-dimensional display space is based on their pair-wise similar-

ity. Similar objects should be placed nearby in the visualization space, dissimilar objects farther apart. In the context of images as objects, suitable similarity measures between images have to be defined [52].
- *Confusion Matrix Visualization:* This technique combines the idea of heatmaps and matrix displays. The classifier confusion matrix (showing the relation between true and predicted classes) is colored according to the value in each cell. The diagonal of the matrix indicates correct classification and all the values other than the diagonal are errors that need to be inspected. Confusion matrix visualizations have been applied to clustering and classification problems in other domains [69].
- *Node-Link Diagrams* are visualizations of (un-)directed graphs [10], in which nodes represents objects and links represent relations between objects.

2.3 Computer Vision Tasks

In the surveyed papers different computer vision tasks were solved by DNNs. These are the following:

- *Classification:* The task is to categorize image pixels into one or more classes.
- *Tracking:* Object tracking is the task of locating moving objects over time.
- *Recognition:* Object recognition is the task of identifying objects in an input image by determining their position and label.
- *Detection:* Given an object and an input image the task in object detection is to localize this object in the image, if it exists.
- *Representation Learning:* This task refers to learning features suitable for object recognition, tracking etc. Examples of such features are points, lines, edges, textures, or geometric shapes.

2.4 Network Architectures

We identified six different types of network architectures in the context of visualization. These types are not mutually exclusive, since all types belong to DNNs, but some architectures are more specific, either w.r.t. the types of layers, the type of connections between the layers or the learning algorithm used.

- *DNN:* Deep Neural Networks are the general type of feed-forward networks with multiple hidden layers.
- *CNN:* Convolutional Neural Networks are a type of feed-forward network specifically designed to mimic the human visual cortex [22]. The architecture consists of multiple layers of smaller neuron collections processing portions of the input image (convolutional layers) generating low-level feature maps. Due to their specific architecture CNNs have much fewer connections and parameters compared to standard DNNs, and thus are easier to train.

- *DCNN:* The Deep Convolution Neural Network is a CNN with a special eight-layer architecture [35]. The first five layers are convolutional layers and the last three layers are fully connected.
- *DBN:* Deep Belief Networks can be seen as a composition of Restricted Boltzmann Machines (RBMs) and are characterized by a specific training algorithm [27]. The top two layers of the network have undirected connections, whereas the lower layers have directed connection with the upper layers.
- *CDBN:* Convolutional Deep Belief Networks are similar to DBNs, containing Convolutional RBMs stacked on one another [38]. Training is performed similarly to DBNs using a greedy layer-wise learning procedure i.e. the weights of trained layers are fixed and considered as input for the next layer.
- *MCDNN:* The Multicolumn Deep Neural Networks is basically a combination of several DNN stacked in column form [6]. The input is processed by all DNNs and their output aggregated to the final output of the DNN.

In the next section we will apply the presented classification scheme (cf. Fig. 1) to the selected papers and provide some statistics on the goals, methods, and application domains. Additionally, we categorize the papers according to the taxonomy of Grün [25] (input modification methods, de-convolutional methods and input reconstruction) if this taxonomy is applicable.

3 Visualizations of Deep Neural Networks

Table 1 provides an overview of all papers included in this survey and their categorization. The table is sorted first by publication year and then by author name. In the following, the collected papers are investigated in detail, whereas the subsections correspond to the categories derived in the previous section.

3.1 Visualization Goals

Table 2 provides an overview of the papers in this category. The most prominent goal is architecture assessment (16 papers). Model quality assessment was covered in 8 and general understanding in 7 papers respectively, while only 3 authors approach interactive integration of user feedback.

Authors who have contributed work on visualizing DNNs with the goal **general understanding** have focused on gaining basic knowledge of how the network performs its task. They aimed to understand what each network layer is doing in general. Most of the work in this category conclude that lower layers of the networks contains representations of simple features like edges and lines, whereas deeper layers tend to be more class-specific and learn complex image features [41, 47, 61]. Some authors developed tools to get a better understanding of learning capabilities

Table 1 Overview of all reviewed papers

Author(s)	Year	Vis. Goal	Vis. Method	CV task	Arch.	Data Sets
Simonyan et al. [61]	2014	General understanding	Pixel Displays	Classification	CNN	ImageNet
Yu et al. [80]	2014	General understanding	Pixel Displays	Classification	CNNs	ImageNet
Li et al. [41]	2015	General understanding	Pixel Displays	Representation learning	DCNN	Buffy Stickmen, ETHZ Stickmen, LSP, Synchronic Activities Stickmen, FLIC, WAF
Montavon et al. [50]	2015	General understanding	Heat maps	Classification	DNNs	ImageNet, MNIST
Yosinski et al. [79]	2015	General understanding	Pixel Displays	Classification	DNN	ImageNet
Mahendran and Vedaldi [47]	2016	General understanding	Pixel Displays	Representation learning	CNN	ILSVRC-2012, VOC2010
Wu et al. [74]	2016	General understanding	Pixel Displays	Recognition	DBN	ChaLearn LAP
Ciresan et al. [6]	2012	Architecture assessment	Pixel Displays, Confusion Matrix	Recognition	MCDNN	MNIST, NIST SD, CASIA-HWDB1.1, GTSRB traffic sign dataset, CIFAR10
Huang [28]	2012	Architecture assessment	Pixel Displays	Representation learning	CDBN	LFW
Szegedy et al. [63]	2013	Architecture assessment	Heat maps	Detection	DNN	VOC2007
Long et al.[45]	2014	Architecture assessment	Pixel Displays	Classification	CNNs	ImageNet, VOC
Taigman et al. [64]	2014	Architecture assessment	Pixel Displays	Representation learning	DNN	SFC, YTF, LFW
Yosinski et al. [78]	2014	Architecture assessment	Pixel Displays	Representation learning	CNN	ImageNet
Zhou et al. [84]	2014	Architecture assessment	Pixel Displays	Recognition	CNN	ImageNet, SUN397, MIT Indoor67, Scene15, SUNAttribute, Caltech-101, Caltech256, Stanford Action40, UIUC Event8
Zhou et al. [82]	2014	Architecture assessment	Pixel Displays	Classification	CNNs	SUN397, Scene15
Mahendran and Vedaldi [46]	2015	Architecture assessment	Pixel Displays	Representation learning	CNN	ILSVRC-2012
Samek et al. [56]	2015	Architecture assessment	Pixel Displays, Heat maps	Classification	DNN	SUN397, ILSVRC-2012, MIT
Wang et al. [70]	2015	Architecture assessment	Pixel Displays	Detection	CNNs	PASCAL3D+
Zhou et al. [83]	2015	Architecture assessment	Heat maps	Recognition	CNNs	ImageNet
Gruen et al. [25]	2016	Architecture assessment	Pixel Displays	Representation learning	DNN	ImageNet
Lin and Maji [42]	2016	Architecture assessment	Pixel Displays	Recognition	CNN	FMD, DTD, KTH-T2b, ImageNet
Nguyen et al. [53]	2016	Architecture assessment	Pixel Displays	Tracking	DNN	ImageNet, ILSVRC-2012
Zintgraf [85]	2016	Architecture assessment	Pixel Displays, HeatMaps	Classification	DCNN	ILSVRC
Erhan et al. [16]	2010	Model quality assessment	Pixel Displays	Representation learning	DBN	MNIST, Caltech-101
Krizhevsky et al. [35]	2012	Model quality assessment	Histogram	Classification	DCNN	ILSVRC-2010, ILSVRC-2012
Dai and Wu [8]	2014	Model quality assessment	Pixel Displays	Classification	CNNs	ImageNet, MNIST
Donahue et al. [11]	2014	Model quality assessment	Similarity layout	Classification	DNN	ILSVRC-2012, SUN397, Caltech-101, Caltech-UCSD Birds
Zeiler and Fergus [81]	2014	Model quality assessment	Pixel Displays, HeatMap	Classification	CNN	ImageNet, Caltech-101, Caltech256
Cao et al. [3]	2015	Model quality assessment	Pixel Displays	Tracking	CNN	ImageNet 2014
Wang et al. [71]	2015	Model quality assessment	Heat maps	Tracking	CNN	ImageNet
Dosovitskiy and Brox [12]	2016	Model quality assessment	Pixel Displays	Representation learning	CNN	ImageNet
Bruckner [2]	2014	User feedback integration	Pixel Displays, Confusion Matrix	Classification	DCNN	CIFAR-10, ILSVRC-2012
Harley [26]	2015	User feedback integration	Pixel Displays, Node-Link-Diagram	Recognition	CNNs	MNIST
Liu et al. [44]	2016	User feedback integration	Pixel Displays, Node-Link-Diagrams	Classification	CNNs	CIFAR10

Table 2 Overview of visualization goals

Category	# Papers	References
Architecture assessment	16	[6, 25, 28, 42, 45, 46, 53, 56, 63, 64, 70, 78, 82–85]
Model quality assessment	8	[3, 8, 11, 12, 16, 35, 71, 81]
General understanding	7	[41, 47, 50, 61, 74, 79, 80]
User feedback integration	3	[2, 26, 44]

of convolutional networks[1] [2, 79]. They demonstrated that such tools can provide a means to visualize the activations produced in response to user inputs and showed how the network behaves on unseen data.

Approaches providing deeper insights into the architecture were placed into the category **architecture assessment**. Authors focused their research on determining how these networks capture representations of texture, color and other features that discriminate an image from another, quite similar image [56]. Other authors tried to assess how these deep architectures arrive at certain decisions [42] and how the input image data affects the decision making capability of these networks under different conditions. These conditions include image scale, object translation, and cluttered background scenes. Further, authors investigated which features are learned, and whether the neurons are able to learn more than one feature in order to arrive at a decision [53]. Also, the contribution of image parts for activation of specific neurons was investigated [85] in order to understand for instance, what part of a dog's face needs to be visible for the network to detect it as a dog. Authors also investigated what types of features are transferred from lower to higher layers [78, 79], and have shown for instance, that scene centric and object centric features are represented differently in the network [84].

Eight papers contributed work on **model quality assessment**. Authors have focused their research on how the individual layers can be effectively visualized, as well as the effect on the network's performance. The contribution of each layer at different levels greatly influence their role played in computer vision tasks. One such work determined how the convolutional layers at various levels of the network show varied properties in tracking purposes [71]. Dosovitskiy and Bronx have shown that higher convolutional layers retain details of object location, color, and contour information of the image [12]. Visualization is used as a means to improve tools for finding good interpretations of features learned in higher levels [16]. Kriszhesvsky et al. focused on performance of individual layers and how performance degrades when certain layers in the network are removed [35].

Some authors researched **user feedback integration**. In the interactive node-link visualization in [26] the user can provide his/her own training data using a drawing area. This method is strongly tied to the used network and training data (MNIST hand written digit). In the *Ml-O-Scope* system users can interactively analyze convolutional neural networks [2]. Users are presented with a visualization of the current model performance, i.e. the a-posteriori probability distribution for input images and Pixel Displays of activations within selected network layers. They are also provided with a user interface for interactive adaption of model hyper-parameters. A visual analytics approach to DNN training has been proposed recently [44]. The authors present 3 case studies in which DNN experts evaluated a network, assessed errors, and found directions for improvement (e.g. adding new layers).

[1]Tools available http://yosinski.com/deepvis and https://github.com/bruckner/deepViz, last accessed 2016-09-08.

Table 3 Overview of visualization methods

Category	Sub-category	# Papers	References
Pixel Displays	Single image	24	[3, 6, 8, 12, 16, 25, 26, 41, 42, 44–47, 53, 56, 61, 70, 71, 74, 78–81, 85]
	Image batch	4	[2, 28, 35, 84]
	Part of image	2	[64, 82]
Heat maps		6	[50, 56, 63, 71, 81, 83, 85]
Confusion matrix		2	[2, 6]
Node-Link-Diagrams		2	[26, 44]
Similarity layout		1	[11]
Histogram		1	[35]

Table 4 Overview of categorization by Grün [25]

Category	# Papers	References
Deconvolution	24	[2, 3, 6, 8, 12, 16, 26, 28, 35, 41, 45, 50, 53, 56, 61, 63, 64, 70, 71, 80, 82–84]
Input modification	6	[44, 74, 78, 79, 81, 85]
Input reconstruction	4	[42, 46, 47, 61]

3.2 Visualization Methods

In this section we describe the different visualization methods applied to DNNs. An overview of the methods is provided in Table 3. We also categorize the papers according to Grün's taxonomy [25] in Table 4. In the following we describe the papers for each visualization method separately.

3.2.1 Pixel Displays

Most of the reviewed work has utilized pixel based activations as a means to visualize different features and layers of deep neural networks. The basic idea behind such visualization is that each pixel represents a data point. The color of the pixel corresponds to an activation value, the maximum gradient w.r.t. to a given class, or a reconstructed image. The different computational approaches for calculating maximum activations, sensitivity values, or reconstructed images are not within the scope of this chapter. We refer to the survey paper for feature visualizations in DNNs [25] and provide a categorization of papers into Grün's taxonomy in Table 4.

Mahendran and Vedaldi [46, 47] have visualized the information contained in the image by using a process of inversion using an optimized gradient descent function. Visualizations are used to show the representations at each layer of the

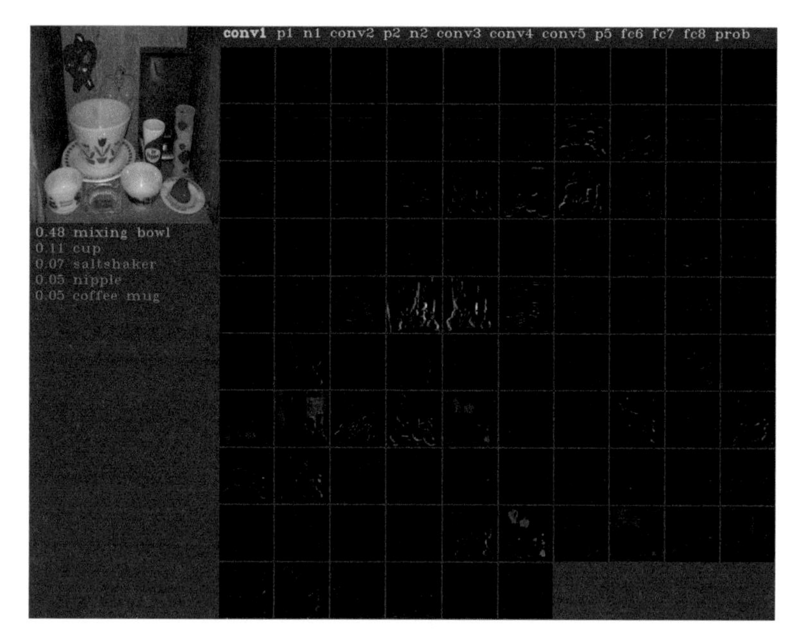

Fig. 2 Pixel based display. Activations of first convolutional layer generated with the DeepVis toolboxfrom [79] https://github.com/yosinski/deep-visualization-toolbox/

network (cf. Fig. 2). All the convolutional layers maintain photographically realistic representations of the image. The first few layers are specific to the input images and form a direct invertible code base. The fully connected layers represent data with less geometry and instance specific information. Activation signals can thus be invert back to images containing parts similar, but not identical to the original images. Cao et al. [3] have used Pixel Displays on complex, cluttered, single images to visualize their results of CNNs with feedback. Nguyen et al. [53] developed an algorithm to demonstrate that single neurons can represent multiple facets. Their visualizations show the type of image features that activate specific neurons. A regularization method is also presented to determine the interpretability of the images to maximize activation. The results suggest that synthesizing visualizations from activated neurons better represent input images in terms of the overall structure and color. Simonyan et al. [61] visualized data for deep convolutional networks. The first visualization is a numerically generated image to maximize a classification score. As second visualization, saliency maps for given pairs of images and classes indicate the influence of pixels from the input image on the respective class score, via back-propagation.

3.2.2 Heat Maps

In most cases, heat maps were used for visualizing the extent of feature activations of specific network layers for various computer vision tasks (e.g. classification [81], tracking [71], detection [83]). Heat maps have also been used to visualize the final network output, e.g. the classifier probability [63, 81]. The heat map visualizations are used to study the contributions of different network layers (e.g. [71]), compare different methods (e.g. [50]), or investigate the DNNs inner features and results on different input images [83]. Zintgraf et al. [85] used heat maps to visualize image regions in favor of, as well as image regions against, a specific class in one image. Authors use different color codings for their heat maps: blue-red-yellow color schemes [71, 81, 83], white-red schemes [50], blue-white-red schemes [85], and also a simple grayscale highlighting interesting regions in white [63].

3.2.3 Confusion Matrix and Histogram

Two authors have shown the confusion matrix to illustrate the performance of the DNN w.r.t., a classification task (see Fig. 3). Bruckner et al. [2] additionally encoded the value in each cell using color (darker color represents higher values). Thus, in this visualization, dark off-diagonal spots correspond to large errors. In [6] the encoding used is different: each cell value is additionally encoded by the size of a square. Cells containing large squares represent large values; a large off-diagonal square corresponds to a large error between two classes. Similarly, in one paper histograms have been used to visualize the decision uncertainty of a classifier, indicating using color whether the highest-probable class is the correct one [35].

3.2.4 Similarity Based Layout

In the context of DNNs, similarity based layouts so far have been applied only by Donahue et al. [11], who specifically used t-distributed stochastic neighbor embedding (t-SNE) [67] of feature representations. The authors projected feature representations of different networks layers into the 2-dimensional space and found

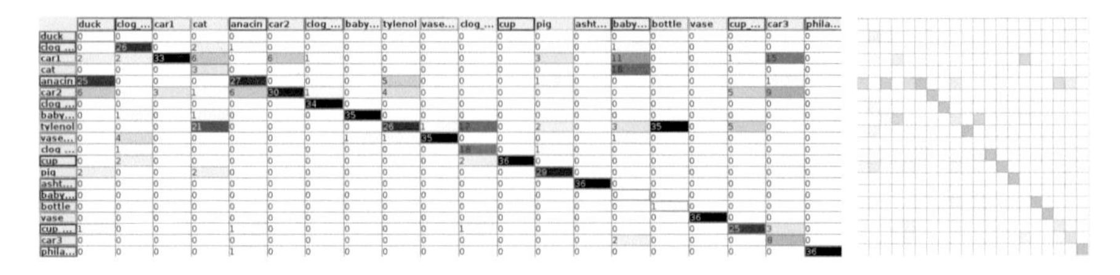

Fig. 3 Confusion Matrix example. Showing classification results for the COIL-20 data set. Screenshots reproduced with software from [59]

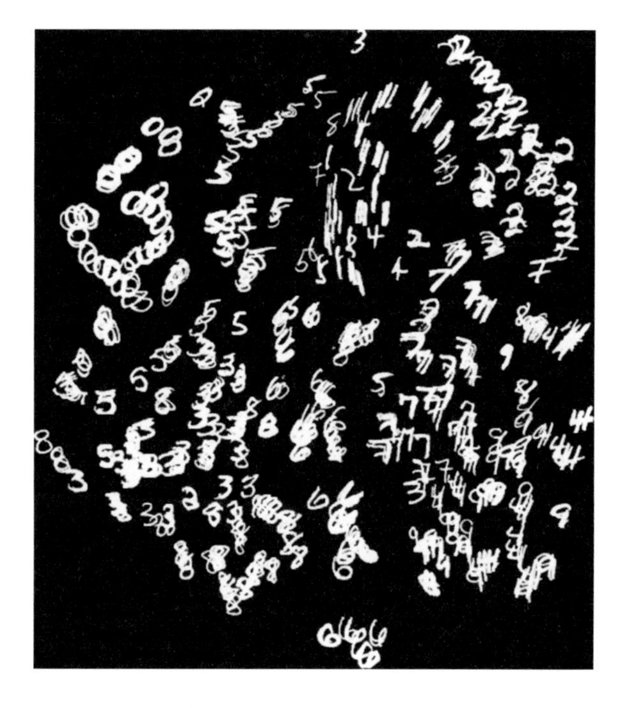

Fig. 4 Similarity based layout of the MNIST data set using raw features. Screenshot was taken with a JavaScript implementation of t-SNE [67] https://scienceai.github.io/tsne-js/

a visible clustering for the higher layers in the network, but none for features of the lower network layer. This finding corresponds to the general knowledge of the community that higher levels learn semantic or high-level features. Further, based on the projection the authors could conclude that some feature representation is a good choice for generalization to other (unseen) classes and how traditional features compare to feature representations learned by deep architectures. Figure 4 provided an example of the latter.

3.2.5 Node-Link Diagrams

Two authors have approach DNN visualization with node-link diagrams (see examples in Fig. 5). In his interactive visualization approach, Adam Harley represented layers in the neural networks as nodes using Pixel Displays, and activation levels as edges [26]. Due to the denseness of connections in DNNs only active edges are visible. Users can draw input images for the network and interactively explore how the DNN is trained. In CNNVis [44] nodes represent neuron clusters and are visualized in different ways (e.g., activations) showing derived features for the clusters.

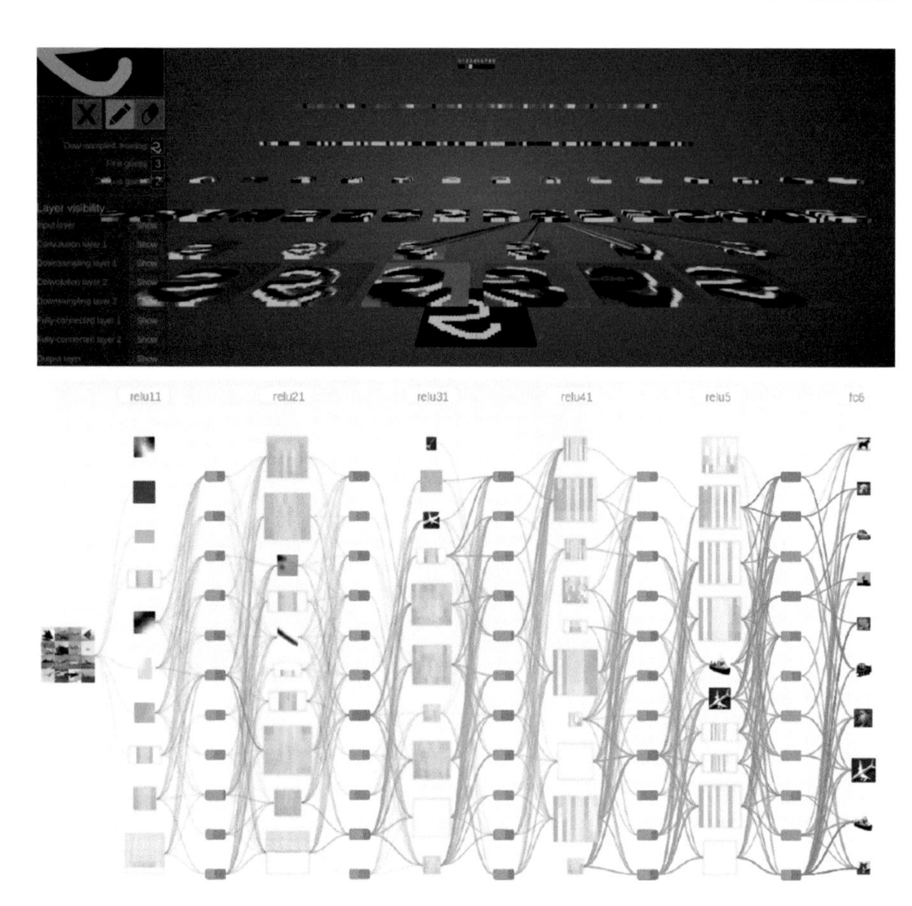

Fig. 5 Node-link diagrams of DNNs. *Top*: Example from [26] taken with the online application at http://scs.ryerson.ca/~aharley/vis/conv/. *Bottom*: screenshot of the CNNVis system [44] taken with the online application at http://shixialiu.com/publications/cnnvis/demo/

3.3 Network Architecture and Computer Vision Task

Table 5 provides a summary of the architecture types. The majority of papers applied visualizations to CNN architectures (18 papers), while 8 papers dealt with the more general case of DNNs. Only 8 papers have investigated more special architectures, like DCNN (4 papers), DBNs (2 papers), CDBN (1 paper) and MCDNNs (1 paper).

Table 6 summarizes the computer vision tasks for which the DNNs have been trained. Most networks were trained for classification (14 papers), some for representation learning and recognition (9 and 6 papers, respectively). Tracking and Detection were pursued the least often.

Table 5 Overview of network architecture types

Category	# Papers	References
CNN	18	[3, 8, 12, 26, 42, 44–47, 61, 70, 71, 78, 80–84]
DNN	8	[11, 25, 50, 53, 56, 63, 64, 79]
DCNN	4	[2, 35, 41, 85]
DBN	2	[16, 74]
CDBN	1	[28]
MCDNN	1	[6]

Table 6 Overview of computer vision tasks

Category	# Papers	References
Classification	14	[2, 8, 11, 35, 44, 45, 50, 56, 61, 79–82, 85]
Representation learning	9	[12, 16, 25, 28, 41, 46, 47, 64, 78]
Recognition	6	[6, 26, 42, 74, 83, 84]
Tracking	3	[3, 53, 71]
Detection	2	[63, 70]

3.4 Data Sets

Table 7 provides an overview of the data sets used in the reviewed papers. In the field of classification and detection, the ImageNet dataset represent the most frequently used dataset, used around 21 times. Other popular datasets used in tasks involving detection and recognition such as Caltech101, Caltech256 etc. have been used 2–3 times (e.g. in [11, 56, 81, 84]).

While ImageNet and its subsets (e.g. ISLVRC) are large datasets with around 10,000,000 images each, there are smaller datasets such as the ETHZ stickmen and VOC2010 which are generally used for fine-grained classification and learning. VOC2010, consisting of about 21,738 images, has been used twice, while more specialized data sets, such as Buffy Stickmen for representation learning, have been used only once in the reviewed papers [41]. There are datasets used in recognition with fewer classes such as CIFAR10, consisting of 60,000 colour images, with about 10 classes; and MNIST used for recognition of handwritten digits.

4 Discussion

In this section we discuss the implications of the findings from the previous section with respect to the research questions. We start the discussion by evaluating the results for the stated research questions.

RQ-1 (Which insights can be gained about DNN models by means of visualization) has been discussed along with the single papers in the previous section in

Table 7 Overview of data sets sorted after their usage. Column "#" refers to the number of papers in this survey using this data set

Data Set	Year	# Images	CV Task	Comment	#	References
ImageNet [9]	2009	14,197,122	Classification, tracking	21,841 synsets	21	[2, 3, 11, 12, 25, 35, 42, 46, 47, 53, 53, 56, 56, 61, 71, 78, 79, 81, 84, 85, 85]
ILSVRC2012 [55]	2015	1,200,000	Classification, detection, representation learning	1000 object categories	7	[2, 11, 35, 46, 47, 53, 56]
VOC2010 [18]	2010	21,738	Detection, representation learning	50/50 train-test split	3	[45, 47, 63]
Caltech-101 [20]	2006	9146	Recognition, classification	101 categories	3	[11, 81, 84]
Places [84]	2014	2,500,000	Classification, recognition	205 scene categories	2	[56, 84]
Sun397 [76]	2010	130,519	Classification, recognition	397 categories	2	[56, 84]
Caltech256 [23]	2007	30,607	Classification,recognition	256 categories	2	[81, 84]
LFW [29]	2007	13,323	Representation learning	5749 faces, 6000 pairs	2	[28, 64]
MNIST [37]	1998	70,000	Recognition	60,000 train, 10,000 test 10 classes, hand-written digits	2	[6, 16]
DTD [5]	2014	5640	Recognition	47 terms (categories)	1	[42]
ChaLearn LAP [17]	2014	47,933	Recognition	RGB-D gesture videos with 249 gestures labels, each 249 for train/testing/validation	1	[74]
SFC [64]	2014	4,400,000	Representation learning	Photos of 4030 people	1	[64]
PASCAL3D+ [75]	2014	30,899	Detection		1	[70]
FLIC [57]	2013	5003	Representation learning	30 movies with person detector, 20% for testing	1	[40]
Synchronic Activities Stick-men [15]	2012	357	Representation learning	Upper-body annotations	1	[40]
Buffy Stickmen [30]	2012	748	Representation learning	Ground-truth stickmen annotations, anno-tated video frames	1	[40]
SUNAttribute [54]	2012	14,000	Recognition	700 categories	1	[84]
CASIA-HWDB1.1 [43]	2011	1,121,749	Recognition	897,758 train, 223,991 test ,3755 classes, Chinese handwriting	1	[6]
GTSRB traffic sign dataset [62]	2011	50,000	Recognition	>40 classes, single-image, multi-class classification	1	[6]
Caltech-UCSD Birds [68]	2011	11,788	Classification	200 categories	1	[11]
YTF [73]	2011	3425	Representation learning	Videos, subset of LFW	1	[64]
Stanford Action40 [77]	2011	9532	Recognition	40 actions, 180–300 images per action class	1	[84]
WAF [14]	2010	525	Representation learning	Downloaded via Google Image Search	1	[40]
LSP [31]	2010	2000	Representation learning	Pose annotated images with 14 joint loca-tion	1	[40]
ETHZ Stickmen [13]	2009	549	Representation learning	Annotated by a 6-part stickman	1	[40]
CIFAR10 [34]	2009	60,000	Recognition	50,000 training and 10,000 test of 10 classes	1	[6]
FMD [60]	2009	1000	Recognition	10 categories, 100 images per category	1	[42]
UIUC Event8 [39]	2007	1579	Recognition	sports event categories	1	[84]
KTH-T2b [4]	2005	4752	Recognition	11 materials captured under controlled scale, pose, and illumination	1	[42]
Scene15 [19]	2005	4485	Recognition	200–400 images per class of 15 class scenes	1	[84]
NIST SD 19 [24]	1995	800,000	Recognition	Forms and digits	1	[6]

detail. We showed by examples which visualizations have previously been shown to lead to which insights. For instance, visualizations are used to learn which features are represented in which layer of a network or which part of the image a certain node reacts to. Additionally, visualizing synthetic input images which maximize activation allows to better understand how a network as a whole works. To strengthen our point here, we additionally provide some quotes from authors:

Heat maps: *"The visualisation method shows which pixels of a specific input image are evidence for or against a node in the network."* [85]

Similarity layout: *"[...] first layers learn 'low-level' features, whereas the latter layers learn semantic or 'high-level' features. [...] GIST or LLC fail to capture the semantic difference [...]"* [11]

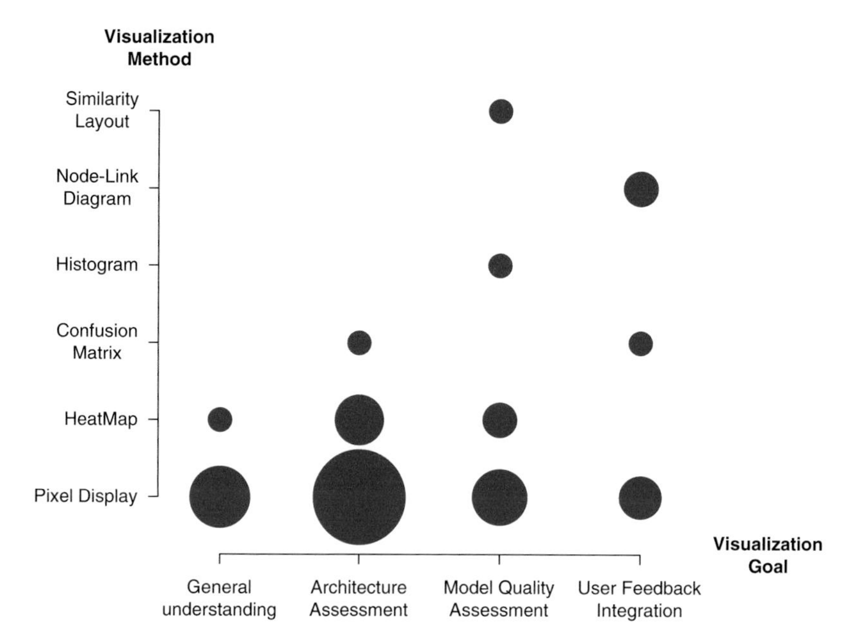

Fig. 6 Relation of visualization goals and applied methods in the surveyed papers following our taxonomy. Size of the circles corresponds to the (square root of the) number of papers in the respective categories. For details on papers see Table 1

Pixel Displays: *"[...] representations on later convolutional layers tend to be somewhat local, where channels correspond to specific, natural parts (e.g. wheels, faces) instead of being dimensions in a completely distributed code. That said, not all features correspond to natural parts [...]"* [79]

The premise to use visualization is thus valid, as the publications agree that visualizations help to understand the functionality and behavior of DNNs in computer vision. This is especially true when investigating specific parts of the DNN.

To answer **RQ-2** (Which visualization methods are appropriate for which kind of insights?) we evaluated which visualizations were applied in the context of which visualization goals. A summary is shown in Fig. 6. It can be seen that not all methods were used in combination with all goals, which is not surprising. For instance, no publication used a similarity layout for assessing the architecture. This provides hints on possibilities for further visualization experiments.

Pixel Displays were prevalent for architecture assessment and general understanding. This is plausible since DNNs for computer vision work on the images themselves. Thus, Pixel Displays preserve the spatial-context of the input data, making the interpretation of the visualization straight-forward. This visualization, however, method has its own disadvantages and might not be the ideal choice in all cases. The visualization design space is extremely limited, i.e. constrained to a simple color mapping. Especially for more complex research questions, extending this space might be worthwhile, as the other visualization examples in this review show.

The fact that a method has not been used w.r.t. a certain goal does not necessarily mean that it would not be appropriate. It merely means that authors so far achieved their goal with a different kind of visualization. The results based on our taxonomy, cf. Fig. 6 and Table 1, hint at corresponding *white spots*. For example, node-link diagrams are well suited to visualize dependencies and relations. Such information could be extracted for architecture assessment as well, depicting which input images and activation levels correlate highly to activations within individual layers of the network. Such a visualization will neither be trivial to create nor to use, since this first three part correlation requires suitable hyper-graph visualization metaphor, but the information basis is promising. Similar example ideas can be constructed for the other white spots in Fig. 6 and beyond.

5 Summary and Conclusion

In this chapter we surveyed visualizations of DNNs in the computer vision domain. Our leading questions were: “Which insights can be gained about DNN models by means of visualization?” and “Which visualization methods are appropriate for which kind of insights?” A taxonomy containing the categories *visualization method*, *visualization goal*, *network architecture type*, *computer vision task*, and *data set* was developed to structure the domain. We found that Pixel Displays were most prominent among the methods, closely followed by heat maps. Both is not surprising, given that images (or image sequences) are the prevalent input data in computer vision. Most of the developed visualizations and/or tools are expert tools, designed for the usage of DNN/computer vision experts. We found no interactive visualization allowing to integrate user feedback directly into the model. The closest approach is the semi-automatic CNNVis tool [44]. An interesting next step would be to investigate which of the methods have been used in other application areas of DNNs, such as speech recognition, where Pixel Displays are not the most straight-forward visualization. It would be also interesting to see which visualization knowledge and techniques could be successfully transferred between these application areas.

References

1. Becker, B., Kohavi, R., Sommerfield, D.: Visualizing the simple Bayesian classifier. In: KDD Workshop Issues in the Integration of Data Mining and Data Visualization (1997)
2. Bruckner, D.M.: Ml-o-scope: a diagnostic visualization system for deep machine learning pipelines. Tech. Rep. UCB/EECS-2014-99, University of California at Berkeley (2014)

3. Cao, C., Liu, X., Yang, Y., Yu, Y., Wang, J., Wang, Z., Huang, Y., Wang, L., Huang, C., Xu, W., Ramanan, D., Huang, T.S.: Look and think twice: capturing top-down visual attention with feedback convolutional neural networks. In: 2015 IEEE International Conference on Computer Vision (ICCV) (2015)
4. Caputo, B., Hayman, E., Mallikarjuna, P.: Class-specific material categorisation. In: Tenth IEEE International Conference on Computer Vision, vol. 1 (2005)
5. Cimpoi, M., Maji, S., Kokkinos, I., Mohamed, S., Vedaldi, A.: Describing textures in the wild. CoRR, abs/1311.3618 (2014)
6. Ciresan, D., Meier, U., Schmidhuber, J.: Multi-column deep neural networks for image classification. In: Computer Vision and Pattern Recognition, pp. 3642–3649 (2012)
7. Cybenko, G.: Approximation by superpositions of a sigmoidal function. Math. Control Signals Syst. **2**(4), 303–314 (1989)
8. Dai, J., Wu, Y.N.: Generative modeling of convolutional neural networks. CoRR, abs/1412.6296 (2014)
9. Deng, J., Dong, W., Socher, R., Li, L.J., Li, K., Fei-Fei, L.: Imagenet: a large-scale hierarchical image database. In: IEEE Conference on Computer Vision and Pattern Recognition, 2009, pp. 248–255, June 2009
10. Di Battista, G., Eades, P., Tamassia, R., Tollis, I.G.: Algorithms for drawing graphs: an annotated bibliography. Comput. Geom. **4**(5), 235–282 (1994)
11. Donahue, J., Jia, Y., Vinyals, O., Hoffman, J., Zhang, N., Tzeng, E., Darrell, T.: Decaf: a deep convolutional activation feature for generic visual recognition. In: International Conference on Machine Learning (2014)
12. Dosovitskiy, A., Brox, T.: Inverting visual representations with convolutional networks. In: IEEE Conference on Computer Vision and Pattern Recognition (2016).
13. Eichner, M., Ferrari, V.: Better appearance models for pictorial structures. In: Proceedings of the British Machine Vision Conference, pp. 3.1–3.11. BMVA Press, Guildford (2009). doi:10.5244/C.23.3
14. Eichner, M., Ferrari, V.: We are family: joint pose estimation of multiple persons. In: Proceedings of the 11th European Conference on Computer Vision: Part I, pp. 228–242. Springer, Berlin/Heidelberg (2010)
15. Eichner, M., Ferrari, V.: Human pose co-estimation and applications. IEEE Trans. Pattern Anal. Mach. Intell. **34**(11), 2282–2288 (2012)
16. Erhan, D., Courville, A., Bengio, Y.: Understanding representations learned in deep architectures. Tech. Rep. 1355, Université de Montréal/DIRO, October 2010
17. Escalera, S., Baró, X., Gonzalez, J., Bautista, M.A., Madadi, M., Reyes, M., Ponce-López, V., Escalante, H.J., Shotton, J., Guyon, I.: Chalearn looking at people challenge 2014: Dataset and results. In: Workshop at the European Conference on Computer Vision (2014)
18. Everingham, M., van Gool, L., Williams, C.K.I., Winn, J., Zisserman, A.: The Pascal visual object classes (voc) challenge. Int. J. Comput. Vis. **88**(2), 303–338 (2010)
19. Fei-Fei, L., Perona, P.: A Bayesian hierarchical model for learning natural scene categories. In: Proceedings of the 2005 IEEE Computer Society Conference on Computer Vision and Pattern Recognition (CVPR'05) - Volume 2 - Volume 02, pp. 524–531. IEEE Computer Society, Washington, DC (2005)
20. Fei-Fei, L., Fergus, R., Perona, P.: One-shot learning of object categories. IEEE Trans. Pattern Anal. Mach. Intell. **28**(4), 594–611 (2006)
21. Fuchs, R., Waser, J., Gröller, E.: Visual human+machine learning. Proc. Vis. 09 **15**(6), 1327–1334 (2009)
22. Fukushima, K., Miyake, S.: Neocognitron: a new algorithm for pattern recognition tolerant of deformations and shifts in position. Pattern Recogn. **15**(6), 455–469 (1982)
23. Griffin, G., Houlub, A., Perona, P.: Caltech-256 object category dataset. Tech. Rep., California Institute of Technology (2007)
24. Grother, P.J.: NIST special database 19 – Handprinted forms and characters database. Technical report, Natl. Inst. Stand. Technol. (NIST) (1995). https://www.nist.gov/sites/default/files/documents/srd/nistsd19.pdf

25. Grün, F., Rupprecht, C., Navab, N., Tombari, F.: A taxonomy and library for visualizing learned features in convolutional neural networks. In: Proceedings of the International Conference on Machine Learning (2016)
26. Harley, A.W.: An Interactive Node-Link Visualization of Convolutional Neural Networks, pp. 867–877. Springer International Publishing, Cham (2015)
27. Hinton, G.E., Osindero, S., Teh, Y.-W.: A fast learning algorithm for deep belief nets. Neural Comput. **18**(7), 1527–1554 (2006)
28. Huang, G.B.: Learning hierarchical representations for face verification with convolutional deep belief networks. In: *Proceedings Conference on Computer Vision and Pattern Recognition*, pp. 2518–2525. IEEE Computer Society, Washington, DC (2012)
29. Huang, G.B., Ramesh, M., Berg, T., Learned-Miller, E.: Labeled faces in the wild: a database for studying face recognition in unconstrained environments. Tech. Rep. 07–49, University of Massachusetts, Amherst, October 2007
30. Jammalamadaka, N., Zisserman, A., Eichner, M., Ferrari, V., Jawahar, C.: Has my algorithm succeeded? an evaluator for human pose estimators. In: European Conference on Computer Vision (2012)
31. Johnson, S., Everingham, M.: Clustered pose and nonlinear appearance models for human pose estimation. In: Proceedings of the British Machine Vision Conference (2010). doi:10.5244/C.24.12
32. Keim, D., Bak, P., Schäfer, M.: Dense Pixel Displays. In: Liu, L., Öszu, M.T. (eds.) Encyclopedia of Database Systems, pp. 789–795. Springer, New York (2009)
33. Kohavi, R.: Data mining and visualization. Invited talk at the National Academy of Engineering US Frontiers of Engineers (NAE) (2000)
34. Krizhevsky, A.: Learning multiple layers of features from tiny images. Tech. Rep., University of Toronto (2009)
35. Krizhevsky, A., Sutskever, I., Hinton, G.E.: Imagenet classification with deep convolutional neural networks. In: Pereira, F., Burges, C.J.C., Bottou, L., Weinberger, K.Q. (eds.) Advances in Neural Information Processing Systems 25, pp. 1097–1105. Curran Associates, Inc., Red Hook (2012)
36. LeCun, Y., Bengio, Y., Hinton, G.: Deep learning. Nature **521** 436–444 (2015)
37. LeCun, Y., Bottou, L., Bengio, Y., Haffner, P.: Gradient-based learning applied to document recognition. Proc. IEEE **86**(11), 2278–2324 (1998)
38. Lee, H., Grosse, R., Ranganath, R., Ng, A.Y.: Convolutional deep belief networks for scalable unsupervised learning of hierarchical representations. In: Proceedings of the 26th Annual International Conference on Machine Learning, pp. 609–616. ACM, New York (2009)
39. Li, L., Fei-Fei, L.: What, where and who? classifying events by scene and object recognition. In: IEEE International Conference on Computer Vision (2007)
40. Li, S., Liu, Z.-Q., Chan, A.B.: Heterogeneous multi-task learning for human pose estimation with deep convolutional neural network. In: Proceedings of the IEEE Conference on Computer Vision and Pattern Recognition Workshops (2014)
41. Li, S., Liu, Z.-Q., Chan, A.B.: Heterogeneous multi-task learning for human pose estimation with deep convolutional neural network. Int. J. Comput. Vis. **113**(1), 19–36 (2015)
42. Lin, T.-Y., Maji, S.: Visualizing and understanding deep texture representations. In: *Conference on Computer Vision and Pattern Recognition* (2016)
43. Liu, C.-L., Yin, F., Wang, D.-H., Wang, Q.-F.: Casia online and offline Chinese handwriting databases. In: 2011 International Conference on Document Analysis and Recognition (2011)
44. Liu, M., Shi, J., Li, Z., Li, C., Zhu, J., Liu, S.: Towards better analysis of deep convolutional neural networks. CoRR, abs/1604.07043 (2016)
45. Long, J., Zhang, N., Darrell, T.: Do convnets learn correspondence? CoRR, abs/1411.1091 (2014)
46. Mahendran, A., Vedaldi, A.: Understanding deep image representations by inverting them. In: Proceedings of the IEEE Conf. on Computer Vision and Pattern Recognition (2015)
47. Mahendran, A., Vedaldi, A.: Visualizing deep convolutional neural networks using natural pre-images. In: International Journal of Computer Vision (2016)

48. McCulloch, W.S., Pitts, W.: A logical calculus of the ideas immanent in nervous activity. Bull. Math. Biophys. **5**(4), 115–133 (1943)
49. Minsky, M., Papert, S.: Perceptrons: An Introduction to Computational Geometry. MIT Press, Cambridge, MA (1969)
50. Montavon, G., Bach, S., Binder, A., Samek, W., Müller, K.-R.: Explaining nonlinear classification decisions with deep Taylor decomposition. CoRR, abs/1512.02479 (2015)
51. Munzner, T.: Visualization Analysis and Design. A K Peters Visualization Series. CRC Press, Boca Raton, FL (2014)
52. Nguyen, G.P., Worring, M.: Interactive access to large image collections using similarity-based visualization. J. Vis. Lang. Comput. **19**(2), 203–224 (2008)
53. Nguyen, A.M., Yosinski, J., Clune, J.: Multifaceted feature visualization: uncovering the different types of features learned by each neuron in deep neural networks. CoRR, abs/1602.03616 (2016)
54. Patterson, G.: Sun attribute database: discovering, annotating, and recognizing scene attributes. In: Proceedings of the 2012 IEEE Conference on Computer Vision and Pattern Recognition (CVPR), pp. 2751–2758. IEEE Computer Society, Washington, DC (2012)
55. Russakovsky, O., Deng, J., Su, H., Krause, J., Satheesh, S., Ma, S., Huang, Z., Karpathy, A., Khosla, A., Bernstein, M., Berg, A.C., Fei-Fei, L.: ImageNet large scale visual recognition challenge. Int. J. Comput. Vis. **115**(3), 211–252 (2015)
56. Samek, W., Binder, A., Montavon, G., Bach, S., Müller, K.-R.: Evaluating the visualization of what a deep neural network has learned. CoRR, abs/1509.06321 (2015)
57. Sapp, B., Taskar, B.: Modec: multimodal decomposable models for human pose estimation. In: Proceedings of the Computer Vision and Pattern Recognition (2013)
58. Seifert, C., Granitzer, M.: User-based active learning. In: Proceedings of 10th International Conference on Data Mining Workshops, pp. 418–425 (2010)
59. Seifert, C., Lex, E.: A novel visualization approach for data-mining-related classification. In: Proceedings of the International Conference on Information Visualisation (IV), pp. 490–495. Wiley, New York (2009)
60. Sharan, L., Rosenholtz, R., Adelson, E.: Material perception: what can you see in a brief glance? J. Vis. **9**, 784 (2009). doi:10.1167/9.8.784
61. Simonyan, K., Vedaldi, A., Zisserman, A.: Deep inside convolutional networks: visualising image classification models and saliency maps. CoRR, abs/1312.6034 (2014)
62. Stallkamp, J., Schlipsing, M., Salmen, J., Igel, C.: The German traffic sign recognition benchmark: a multi-class classification competition. In: Neural Networks (IJCNN), The 2011 International Joint Conference on (2011)
63. Szegedy, C., Toshev, A., Erhan, D.: Deep neural networks for object detection. In: Burges, C.J.C., Bottou, L., Welling, M., Ghahramani, Z., Weinberger, K.Q. (eds.) Advances in Neural Information Processing Systems 26, pp. 2553–2561. Curran Associates, Inc., Red Hook (2013)
64. Taigman, Y., Yang, M., Ranzato, M., Wolf, L.: Deepface: closing the gap to human-level performance in face verification. In: 2014 IEEE Conference on Computer Vision and Pattern Recognition, pp. 1701–1708 (2014)
65. Thearling, K., Becker, B., DeCoste, D., Mawby, W., Pilote, M., Sommerfield, D.: Chapter Visualizing data mining models. In: Information Visualization in Data Mining and Knowledge Discovery, pp. 205–222. Morgan Kaufmann Publishers Inc., San Francisco, CA (2001)
66. Urbanek, S.: Exploring statistical forests. In: Proceedings of the 2002 Joint Statistical Meeting (2002)
67. van der Maaten, L., Hinton, G.E.: Visualizing high-dimensional data using t-sne. J. Mach. Learn. Res. **9**, 2579–2605 (2008)
68. Wah, C., Branson, S., Welinder, P., Perona, P., Belongie, S.: The caltech-ucsd birds-200-2011 dataset. Tech. Rep. CNS-TR-2011-001, California Institute of Technology (2011)
69. Wang, J., Yu, B., Gasser, L.: Classification visualization with shaded similarity matrices. Tech. Rep., GSLIS University of Illinois at Urbana-Champaign (2002)
70. Wang, J., Zhang, Z., Premachandran, V., Yuille, A.L.: Discovering internal representations from object-cnns using population encoding. CoRR, abs/1511.06855 (2015)

71. Wang, L., Ouyang, W., Wang, X., Lu, H.: Visual tracking with fully convolutional networks. In: IEEE International Conference on Computer Vision (2015)
72. Wilkinson, L., Friendly, M.: The history of the cluster heat map. Am. Stat. **63**(2), 179–184 (2009)
73. Wolf, L., Hassner, T., Maoz, I.: Face recognition in unconstrained videos with matched background similarity. In: Proceedings of the IEEE Conference on Computer Vision and Pattern Recognition (2011)
74. Wu, D., Pigou, L., Kindermans, P.J., Le, N.D.H., Shao, L., Dambre, J., Odobez, J.M.: Deep dynamic neural networks for multimodal gesture segmentation and recognition. IEEE Trans. Pattern Anal. Mach. Intell. **38**(8), 1583–1597 (2016)
75. Xiang, Y., Mottaghi, R., Savarese, S.: Beyond Pascal: a benchmark for 3d object detection in the wild. In: IEEE Winter Conference on Applications of Computer Vision (2014)
76. Xiao, J., Hays, J., Ehinger, K.A., Oliva, A., Torralba, A.: Sun database: large-scale scene recognition from abbey to zoo. In: 2010 IEEE Conference on Computer Vision and Pattern Recognition (2010)
77. Yao, B., Jiang, X., Khosla, A., Lin, A.L., Guibas, L., Fei-Fei, L.: Human action recognition by learning bases of action attributes and parts. In: International Conference on Computer Vision (2011)
78. Yosinski, J., Clune, J., Bengio, Y., Lipson, H.: How transferable are features in deep neural networks? In: Ghahramani, Z., Welling, M., Cortes, C., Lawrence, N.D., Weinberger, K.Q. (eds.) Advances in Neural Information Processing Systems 27, pp. 3320–3328. Curran Associates, Inc., Red Hook (2014)
79. Yosinski, J., Clune, J., Nguyen, A., Fuchs, T., Lipson, H.: Understanding neural networks through deep visualization. In: Proceedings of the International Conference on Machine Learning (2015)
80. Yu, W., Yang, K., Bai, Y., Yao, H., Rui, Y.: Visualizing and comparing convolutional neural networks. CoRR, abs/1412.6631 (2014)
81. Zeiler, M.D., Fergus, R.: Visualizing and understanding convolutional networks. In: Computer Vision 13th European Conference (2014)
82. Zhou, B., Khosla, A., Lapedriza, À., Oliva, A., Torralba, A.: Object detectors emerge in deep scene cnns. CoRR, abs/1412.6856 (2014)
83. Zhou, B., Khosla, A., Lapedriza, À., Oliva, A., Torralba, A.: Learning deep features for discriminative localization. CoRR, abs/1512.04150 (2015)
84. Zhou, B., Lapedriza, À., Xiao, J., Torralba, A., Oliva, A.: Learning deep features for scene recognition using places database. In: Ghahramani, Z., Welling, M., Cortes, C., Lawrence, N.D., Weinberger, K.Q. (eds.) Advances in Neural Information Processing Systems 27, pp. 487–495. Curran Associates, Inc., Red Hook (2014)
85. Zintgraf, L.M., Cohen, T., Welling, M.: A new method to visualize deep neural networks. CoRR, abs/1603.02518 (2016)

Part III
Regulatory Solutions

Beyond the EULA: Improving Consent for Data Mining

Luke Hutton and Tristan Henderson

Abstract Companies and academic researchers may collect, process, and distribute large quantities of personal data without the explicit knowledge or consent of the individuals to whom the data pertains. Existing forms of consent often fail to be appropriately readable and ethical oversight of data mining may not be sufficient. This raises the question of whether existing consent instruments are sufficient, logistically feasible, or even necessary, for data mining. In this chapter, we review the data collection and mining landscape, including commercial and academic activities, and the relevant data protection concerns, to determine the types of consent instruments used. Using three case studies, we use the new paradigm of human-data interaction to examine whether these existing approaches are appropriate. We then introduce an approach to consent that has been empirically demonstrated to improve on the state of the art and deliver meaningful consent. Finally, we propose some best practices for data collectors to ensure their data mining activities do not violate the expectations of the people to whom the data relate.

Abbreviations

AKI Acute kidney injury
An abrupt loss of kidney function.

EU European Union
A political union of countries primarily located within Europe.

EULA End User License Agreement
A contract between the provider of software or services and its users, establishing the terms under which it can be used.

L. Hutton (✉)
Centre for Research in Computing, The Open University, Milton Keynes MK7 6AA, UK
e-mail: luke.hutton@open.ac.uk

T. Henderson
School of Computer Science, University of St Andrews, St Andrews KY16 9SX, UK
e-mail: tnhh@st-andrews.ac.uk

T. Cerquitelli et al. (eds.), *Transparent Data Mining for Big and Small Data*,
Studies in Big Data 32, DOI 10.1007/978-3-319-54024-5_7

GDPR General Data Protection Regulation
The new European Union data protection regulation, which governs how personal data can be used within the EU and exported outside of the EU.

HCI Human-computer interaction
A discipline of computer science investigating the relationship between computer systems and their users.

HDI Human-data interaction
An emerging discipline of computer science concerned with the relationship between individuals and the increasing range of data flows and inferences that are created and made about them.

IRB Institutional review board
A committee that reviews and approves research studies involving humans.

NHS National Health Service
The universal health care system of the United Kingdom.

SUS Secondary Uses Service
A database of previous NHS treatments that can be made available to researchers.

T3 Tastes, Ties and Time
A 2006 dataset of Facebook profiles collected by researchers at Harvard.

1 Introduction

The ability of companies to collect, process, and distribute large quantities of personal data, and to further analyse, mine and generate new data based on inferences from these data, is often done without the explicit knowledge or consent of the individuals to whom the data pertains. Consent instruments such as privacy notices or End User License Agreements (EULAs) are widely deployed, often presenting individuals with thousands of words of legal jargon that they may not read nor comprehend, before soliciting agreement in order to make use of a service. Indeed, even if an individual does have a reasonable understanding of the terms to which they have agreed, such terms are often carefully designed to extend as much flexibility to the data collector as possible to obtain even more data, distribute them to more stakeholders, and make inferences by linking data from multiple sources, despite no obvious agreement to these new practices.

The lack of transparency behind data collection and mining practices threatens the agency and privacy of data subjects, with no practical way to control these invisible data flows, nor correct misinformation or inaccurate and inappropriate inferences derived from linked data. Existing data protection regimes are often insufficient as they are predicated on the assumption that an individual is able to detect when a data protection violation has occurred in order to demand recourse, which is rarely the case when data are opaquely mined at scale.

These challenges are not unique to commercial activities, however. Academic researchers often make use of datasets containing personal information, such as those collected from social network sites or devices such as mobile phones or fitness trackers. Most researchers are bound by an obligation to seek ethical approval from an institutional review board (IRB) before conducting their research. The ethical protocols used, however, are inherited from post-war concerns regarding biomedical experiments, and may not be appropriate for Internet-mediated research, where millions of data points can be collected without any personal interventions. This raises the question of whether existing consent instruments are sufficient, logistically feasible, or even necessary, for research of this nature.

In this chapter we first review the data collection and mining landscape, including commercial and academic activities, and the relevant data protection laws, to determine the types of consent instruments used. Employing the newly-proposed paradigm of Human-Data Interaction, we examine three case studies to determine whether these mechanisms are sufficient to uphold the expectations of individuals, to provide them with sufficient agency, legibility and negotiability, and whether privacy norms are violated by secondary uses of data which are not explicitly sanctioned by individuals. We then discuss various new dynamic and contextual approaches to consent, which have been empirically demonstrated to improve on the state of the art and deliver meaningful consent. Finally, we propose some best practices that data collectors can adopt to ensure their data mining activities do not violate the expectations of the people to whom the data relate.

2 Background

Data mining is the statistical analysis of large-scale datasets to extract additional patterns and trends [15]. This has allowed commercial, state, and academic actors to answer questions which have not previously been possible, due to insufficient data, analytical techniques, or computational power. Data mining is often characterised by the use of aggregate data to identify traits and trends which allow the identification and characterisation of clusters of people rather than individuals, associations between events, and forecasting of future events. As such, it has been used in a number of real-world scenarios such as optimising the layout of retail stores, attempts to identify disease trends, and mass surveillance. Many classical data mining and knowledge discovery applications involve businesses or marketing [11], such as clustering consumers into groups and attempting to predict their behaviour. This may allow a business to understand their customers and target promotions appropriately. Such profiling can, however, be used to characterise individuals for the purpose of denying service when extending credit, leasing a property, or acquiring insurance. In such cases, the collection and processing of sensitive data can be invasive, with significant implications for the individual, particularly where decisions are made on the basis of inferences that may not be accurate, and to which

Table 1 Some relevant EU legislation that may apply to data mining activities

Legislation	Some relevant sections
Data Protection Directive 95/46/EC	Data processing (Art 1), fair processing (Art 6(1)), purpose limitation (Art 6(2)), proportionality (Art 6(3)), consent (Art 7(1)), sensitive data (Art 8)
E-Privacy Directive 2002/58/EC as amended by 2009/136/EC	Cookies (Art 5), traffic data (Art 6), location data (Art 9), unsolicited communication (Art 13)
General Data Protection Regulation 2016/679	Consent (Art 7), right to be forgotten (Art 17), right to explanation (Art 22), privacy by design (Art 25)

the individual is given no right of reply. This has become more important of late, as more recent data mining applications involve the analysis of personal data, much of which is collected by individuals and contributed to marketers in what has been termed "self-surveillance" [24]. Such personal data have been demonstrated to be highly valuable [48], and have even been described as the new "oil" in terms of the value of their resource [55]. Value aside, such data introduce new challenges for consent as they can often be combined to create new inferences and profiles where previously data would have been absent [16].

Data mining activities are legitimised through a combination of legal and self-regulatory behaviours. In the European Union, the Data Protection Directive [10], and the forthcoming General Data Protection Regulation (GDPR) that will succeed it in 2018 [43] govern how data mining can be conducted legitimately. The e-Privacy Directive also further regulates some specific aspects of data mining such as cookies (Table 1). In the United States, a self-regulatory approach is generally preferred, with the Federal Trade Commission offering guidance regarding privacy protections [41], consisting of six core principles, but lacking the coverage or legal backing of the EU's approach.

Under the GDPR, the processing of personal data for any purpose, including data mining, is subject to explicit opt-in consent from an individual, prior to which the data controller must explicitly state what data are collected, the purpose of processing them, and the identity of any other recipients of the data. Although there are a number of exceptions, consent must generally be sought for individual processing activities, and cannot be broadly acquired *a priori* for undefined future uses, and there are particular issues with data mining: transparency and accountability [7]. Solove [47] acknowledges these regulatory challenges, arguing that paternalistic approaches are not appropriate, as these deny people the freedom to consent to particular beneficial uses of their data. The timing of consent requests and the focus of these requests need to be managed carefully; such thinking has also become apparent in the GDPR.[1] The call for dynamic consent is consistent with Nissenbaum's model of contextual integrity [38], which posits that all information exchanges are subject to context-specific norms, which governs to whom and for what purpose information sharing can be considered appropriate. When the context is disrupted, perhaps by changing with whom data are shared, or for what purpose, privacy violations can occur when this is not consistent with the norms of the existing context. Therefore, consent can help to uphold contextual integrity by ensuring that if the context is perturbed, consent is renegotiated, rather than assumed.

Reasoning about how personal data are used has resulted in a new paradigm, *human-data interaction*, which places humans at the centre of data flows and provides a framework for studying personal data collection and use according to three themes [34]:

- *Legibility*: Often, data owners are not aware that data mining is even taking place. Even if they are, they may not know what is being collected or analysed, the purpose of the analysis, or the insights derived from it.
- *Agency*: The opaque nature of data mining often denies data owners agency. Without any engagement in the practice, people have no ability to provide meaningful consent, if they are asked to give consent at all, nor correct flawed data or review inferences made based on their data.
- *Negotiability*: The context in which data are collected and processed can often change, whether through an evolving legislative landscape, data being traded between organisations, or through companies unilaterally changing their privacy policies or practices. Analysis can be based on the linking of datasets derived from different stakeholders, allowing insights that no single provider could make. This is routinely the case in profiling activities such as credit scoring. Even where individuals attempt to obfuscate their data to subvert this practice, it is often possible to re-identify them from such linked data [37]. Data owners should have the ability to review how their data are used as circumstances change in order to uphold contextual integrity.

Early data protection regulation in the 1980s addressed the increase in electronic data storage and strengthened protections against unsolicited direct marketing [49]. Mail order companies were able to develop large databases of customer details to enable direct marketing, or the trading of such information between companies. When acquiring consent for the processing of such information became mandatory, such as under the 1984 Data Protection Act in the UK, this generally took the form of a checkbox on paper forms, where a potential customer could indicate their willingness for secondary processing of their data. As technology has evolved away from mail-in forms being the primary means of acquiring personal information, and the scope and intent of data protection moves from regulating direct marketing to

[1]For example, Article 7(3) which allows consent to be withdrawn, and Article 17 on the "right to be forgotten" which allows inferences and data to be erased.

a vast range of data-processing activities, there has been little regulatory attention paid to how consent is acquired. As such, consent is often acquired by asking a user to tick a checkbox to opt-in or out of secondary use of their data. This practice is well-entrenched, where people are routinely asked to agree to an End-User Licence Agreement (EULA) before accessing software, and multiple terms of service and privacy policies before accessing online services, generally consisting of a long legal agreement and an "I Agree" button.

A significant body of research concludes that such approaches to acquiring consent are flawed. Luger et al. find that the terms and conditions provided by major energy companies are not sufficiently readable, excluding many from being able to make informed decisions about whether they agree to such terms [29]. Indeed, Obar and Oeldorf-Hirsch find that the vast majority of people do not even read such documents [39], with all participants in a user study accepting terms including handing over their first-born child to use a social network site. McDonald and Cranor measure the economic cost of reading lengthy policies [30], noting the inequity of expecting people to spend an average of 10 min of their time reading and comprehending a complex document in order to use a service. Freidman et al. caution that simply including more information and more frequent consent interventions can be counter-productive, by frustrating people and leading them to making more complacent consent decisions [12].

Academic data mining is subject to a different regulatory regime, with fewer constraints over the secondary use of data from a data protection perspective. This is balanced by an ethical review regime, rooted in post-war concern over a lack of ethical rigour in biomedical research. In the US, ethical review for human subjects research via an institutional review board (IRB) is necessary to receive federal funding, and the situation is similar in many other countries. One of the central tenets of ethical human research is to acquire informed consent before a study begins [5]. As such, institutions have developed largely standardised consent instruments [1] which researchers can use to meet these requirements. While in traditional lab-based studies, these consent procedures can be accompanied by an explanation of the study from a researcher, or the opportunity for a participant to ask any questions, this affordance is generally not available in online contexts, effectively regressing to the flawed EULAs discussed earlier.

Some of these weaknesses have been examined in the literature. Hamnes et al. find that consent documents in rheumatological studies are not sufficiently readable for the majority of the population [14], a finding which is supported by Vučemilo and Borovečki who also find that medical consent forms often exclude important information [53]. Donovan-Kicken et al. examine the sources of confusion when reviewing such documents [8], which include insufficient discussion of risk and lengthy or overly complex language. Munteanu et al. examine the ethics approval process in a number of HCI research case studies, finding that participants often agreed to consent instruments they have not read or understood, and the rigidity of such processes can often be at odds with such studies where a "situational interpretation" of an agreed protocol is needed [35]. There also lacks agreement among researchers about how to conduct such research in an ethical manner, with Vitak et al. finding particular variability regarding whether data should be collected at large scale without consent, or if acquiring consent in such cases is even possible [52].

Existing means of acquiring consent are inherited from a time when the scope of data collection and processing was perhaps constrained and could be well understood. Now, even when the terms of data collection and processing are understood as written, whether registering for an online service, or participating in academic research, it is not clear that the form of gaining the consent was meaningful, or sufficient. Someone may provide consent to secondary use of their data, without knowing what data this constitutes, who will be able to acquire it, for what purpose, or when. This is already a concern when considering the redistribution and processing of self-disclosed personally identifiable information, but becomes increasingly complex when extended to historical location data, shopping behaviours, or social network data, much of which are not directly provided by the individual, and are nebulous in scale and content. Moreover concerns may change over time (the so-called "privacy paradox" [3] that has been demonstrated empirically [2, 4]), which may require changes to previously-granted consent.

Returning to our three themes of *legibility*, *agency*, and *negotiability*, we can see that:

- Existing EULAs and consent forms may not meet a basic standard of *legibility*, alienating significant areas of the population from understanding what they are being asked to agree to. Furthermore, the specific secondary uses of their data are often not explained.
- EULAs and consent forms are often only used to secure permission once, then often never again, denying people *agency* to revoke their consent when a material change in how their data are used arises.
- Individuals have no power to meaningfully *negotiate* how their data are used, nor to intelligently adopt privacy-preserving behaviours, as they generally do not know which data attributed to them is potentially risky.

3 Case Studies

In this section we examine a number of real-world case studies to identify instances where insufficient consent mechanisms were employed, failing to provide people with legibility, agency, and negotiability.

3.1 Taste, Ties, and Time

In 2006, researchers at Harvard University collected a dataset of Facebook profiles from a cohort of undergraduate students, named "Tastes, Ties and Time" (T3) [27]. At the time, Facebook considered individual universities to comprise networks where members of the institution could access the full content of each other's profiles, despite not having an explicit friendship with each other on the service. This design was exploited by having research assistants at the same institution manually extract the profiles of each member of the cohort.

Subsequently, an anonymised version of the dataset was made publicly available, with student names and identifiers removed, and terms and conditions for downloading the dataset made it clear that deanonymisation was not permitted. Unfortunately, this proved insufficient, with aggregate statistics such as the size of the cohort making it possible to infer the college the dataset was derived from, and as some demographic attributes were only represented by a single student, it was likely that individuals could be identified [56].

Individuals were not aware that the data collection took place, and did not consent to its collection, processing, nor subsequent release. As such, this case falls short in our themes for acceptable data-handling practices:

- **Legibility**: Individuals were not aware their data was collected or subsequently released. With a tangible risk of individuals being identified without their knowledge, the individual is not in a position to explore any legal remedies to hold Facebook or the researchers responsible for any resulting harms. In addition, even if consent were sought, it can be difficult for individuals to conceptualise exactly which of their data would be included, considering the large numbers of photos, location traces, status updates, and biographical information a typical user might accrue over years, without an accessible means of visualising or selectively disclosing these data.
- **Agency**: Without notification, individual users had no way to opt-out of the data collection, nor prevent the release of their data. As a side-effect of Facebook's university-only network structure at the time, the only way for somebody to avoid their data being used in such a manner was to leave these institution networks, losing much of the utility of the service in the process. This parallels Facebook's approach to releasing other products, such as the introduction of News Feed in 2006. By broadcasting profile updates to one's network, the effective visibility of their data was substantially increased, with no way to opt-out of the feature without leaving the service entirely. This illusory loss of control was widely criticised at the time [19].
- **Negotiability**: In this respect, the user's relationship with Facebook itself is significant. In addition to IRB approval, the study was conducted with Facebook's permission [27], but Facebook's privacy policy at the time did not allow for Facebook to share their data with the researchers.[2] Therefore, the existing context for sharing information on Facebook was disrupted by this study. This includes the normative expectation that data are shared with Facebook for the purpose of sharing information with one's social network, and not myriad third parties. In addition, no controls were extended to the people involved to prevent it from happening, or to make a positive decision to permit this new information-sharing context.

[2]Facebook Privacy Policy, February 2006: http://web.archive.org/web/20060406105119/http://www.facebook.com/policy.php.

3.2 Facebook Emotional Contagion Experiment

In 2012, researchers at Facebook and Cornell University conducted a large-scale experiment on 689,003 Facebook users. The study manipulated the presentation of stories in Facebook's News Feed product, which aggregates recent content published by a user's social network, to determine whether biasing the emotional content of the news feed affected the emotions that people expressed in their own disclosures [26].

While the T3 study highlighted privacy risks of nonconsensual data sharing, the emotional contagion experiment raises different personal risks from inappropriate data mining activities. For example, for a person suffering from depression, being subjected to a news feed of predominantly depressive-indicative content could have catastrophic consequences, particularly considering the hypothesis of the experiment that depressive behaviour would increase under these circumstances. Considering the scale at which the experiment was conducted, there was no mechanism for excluding such vulnerable people, nor measuring the impact on individuals to mitigate such harms. Furthermore, as the study was not age-restricted, children may have unwittingly been subjected to the study [18]. Rucuber notes that the harms to any one individual in such experiments can be masked by the scale of the experiment [42].

Beyond the research context, this case highlights the broader implications of the visibility of media, whether socially-derived or from mainstream media, being algorithmically controlled. Napoli argues that this experiment highlights Facebook's ability to shape public discourse by altering the news feed's algorithm to introduce political bias [36], without any governance to ensure that such new media are acting in the public interest. The majority of Facebook users do not know that such filtering happens at all, and the selective presentation of content from one's social network can cause social repercussions where the perception is that individuals are withholding posts from someone, rather than an algorithmic intervention by Facebook [9].

This case shows one of the greater risks of opaque data mining. Where people are unaware such activities are taking place, they lose all power to act autonomously to minimise the risk to themselves, even putting aside the responsibility of the researchers in this instance. We now consider how this case meets our three core themes:

- **Legibility**: Individuals were unaware that they were participants in the research. They would have no knowledge or understanding of the algorithms which choose which content is presented on the news feed, and how they were altered for this experiment, nor that the news feed is anything other than a chronological collection of content provided by their social network. Without this insight, the cause of a perceptible change in the emotional bias in the news feed can not be reasoned. Even if one is aware that the news feed is algorithmically controlled, without knowing which data are collected or used in order to determine the relevance of individual stories, it is difficult to reason why certain stories are displayed.

- **Agency**: As in the T3 case, without awareness of the experiment being conducted, individuals were unable to provide consent, nor opt-out of the study. Without an understanding of the algorithms which drive the news feed, nor how they were adjusted for the purposes of this experiment, individuals are unable to take actions, such as choosing which information to disclose or hide from Facebook in an effort to control the inferences Facebook makes, nor to correct any inaccurate inferences. At the most innocuous level, this might be where Facebook has falsely inferred a hobby or interest, and shows more content relating to that. Of greater concern is when Facebook, or the researchers in this study, are unable to detect when showing more depressive-indicative content could present a risk.
- **Negotiability**: In conducting this study, Facebook unilaterally changed the relationship its users have with the service, exploiting those who are unable to control how their information is used [45]. At the time of the study, the Terms of Service to which users agree when joining Facebook did not indicate that data could be used for research purposes [18], a clause which was added after the data were collected. As a commercial operator collaborating with academic researchers, the nature of the study was ambiguous, with Facebook having an internal product improvement motivation, and Cornell researchers aiming to contribute to generalisable knowledge. Cornell's IRB deemed that they did not need to review the study because Facebook provided the data,[3] but the ethical impact on the unwitting participants is not dependent on who collected the data. As Facebook has no legal requirement to conduct an ethical review of their own research, and without oversight from the academic collaborators, these issues did not surface earlier. Facebook has since adopted an internal ethics review process [23], however it makes little reference to mitigating the impact on participants, and mostly aims to maximise benefit to Facebook. Ultimately, these actions by researchers and institutions with which individuals have no prior relationship serve to disrupt the existing contextual norms concerning people's relationship with Facebook, without extending any ability to renegotiate this relationship.

3.3 NHS Sharing Data with Google

In February 2016, the Google subsidiary DeepMind announced a collaboration with the National Health Service's Royal Free London Trust to build a mobile application titled Streams to support the detection of acute kidney injury (AKI) using machine learning techniques. The information sharing agreement permitting this collaboration gives DeepMind ongoing and historical access to all identifiable patient data collected by the trust's three hospitals [21].

[3]Cornell statement: https://perma.cc/JQ2L-TEXQ.

While the project is targeted at supporting those at risk of AKI, data relating to all patients are shared with DeepMind, whether they are at risk or not. There is no attempt to gain the consent of those patients, or to provide an obvious opt-out mechanism. The trust's privacy policy only allows data to be shared without explicit consent for "your direct care".[4] Considering that Streams is only relevant to those being tested for kidney disease, it follows that for most people, their data are collected and processed without any direct care benefit [20], in violation of this policy. Given the diagnostic purpose of the app, such an application could constitute a medical device, however no regulatory approval was sought by DeepMind or the trust [20].

Permitting private companies to conduct data mining within the medical domain disrupts existing norms, by occupying a space that lies between direct patient care and academic research. Existing ethical approval and data-sharing regulatory mechanisms have not been employed, or are unsuitable for properly evaluating the potential impacts of such work. By not limiting the scope of the data collection nor acquiring informed consent, there is no opportunity for individuals to protect their data. In addition, without greater awareness of the collaboration, broader public debate about the acceptability of the practice is avoided, which is of importance considering the sensitivity of the data involved. Furthermore, this fairly limited collaboration can normalise a broader sharing of data in the future, an eventuality which is more likely given an ongoing strategic partnership forged between DeepMind and the trust [20]. We now consider this case from the perspective of our three themes:

- **Legibility**: Neither patients who could directly benefit from improved detection of AKI, nor all other hospital patients, were aware that their data were being shared with DeepMind. Indeed, this practice in many data mining activities—identifying patterns to produce insight from myriad data—risks violating a fundamental principle of data protection regulation; that of proportionality [43].
- **Agency**: The NHS collects data from its functions in a number of databases, such as the Secondary Uses Service which provides a historical record of treatments in the UK, and can be made available to researchers. Awareness of this database and its research purpose is mostly constrained to leaflets and posters situated in GP practices. If patients wish to opt-out of their data being used they must insist on it, and are likely to be reminded of the public health benefit and discouraged from opting-out [6]. Without being able to assume awareness of the SUS, nor individual consent being acquired, it is difficult for individuals to act with agency. Even assuming knowledge of this collaboration, it would require particular understanding of the functions of the NHS to know that opting-out of the SUS would limit historical treatment data made available to DeepMind. Even where someone is willing to share their data to support their direct care,

[4]Royal Free London Trust Privacy Statement: https://perma.cc/33YE-LYPF.

Table 2 Summary of how the three case studies we examine violate the principles of human-data interaction

Case study	HDI principle	How was principle violated?
Taste, Ties and Time	Legibility	Individuals unaware of data collection
	Agency	No way to opt-out of data collection
	Negotiability	Data collection violated normative expectation with Facebook
Facebook emotional contagion experiment	Legibility	Individuals unaware they were participants
	Agency	No way to opt-out of participation
	Negotiability	Research not permitted by Facebook's terms
NHS sharing data with Google	Legibility	Patients unaware of data sharing
	Agency	Poor awareness of secondary uses of data and difficulty of opting out
	Negotiability	Data mining can violate normative expectations of medical confidentiality

they may wish to redact information relating to particularly sensitive diagnoses or treatments, but have no mechanism to do so.

- **Negotiability**: The relationship between patients and their clinicians is embodied in complex normative expectations of confidentiality which are highly context-dependent [44]. Public understanding of individual studies is already low [6], and the introduction of sophisticated data mining techniques into the diagnostic process which existing regulatory mechanisms are not prepared for disrupts existing norms around confidentiality and data sharing. The principle of negotiability holds that patients should be able to review their willingness to share data as their context changes, or the context in which the data are used. Existing institutions are unable to uphold this, and the solution may lie in increased public awareness and debate, and review of policy and regulatory oversight to reason a more appropriate set of norms.

How each of these case studies meets the principles of legibility, agency, and negotiability is summarised in Table 2.

4 Alternative Consent Models

In Sect. 2 we discussed some of the shortcomings with existing means of acquiring consent for academic research and commercial services including data mining, and discussed three case studies in Sect. 3. Many of the concerns in these case studies revolved around an inability to provide or enable consent on the part of participants. We now discuss the state-of-the-art in providing meaningful consent for today's data-mining activities.

The acquisition of informed consent can broadly be considered to be *secured* or *sustained* in nature [28]. Secured consent encompasses the forms we discussed in Sect. 2, where consent is gated by a single EULA or consent form at the beginning of the data collection process and not revisited. Conversely, sustained consent involves ongoing reacquisition of consent over the period that the data are collected or used. This might mean revisiting consent when the purpose of the data collection or processing has changed, such as if data are to be shared with different third parties, or if the data subject's context has changed. Each interaction can be viewed as an individual consent *transaction* [31]. In research, this can also mean extending more granular control to participants over which of their data are collected, such as in Sleeper et al.'s study into self-censorship behaviours on Facebook, where participants could choose which status updates they were willing to share with researchers [46]. This approach has a number of advantages. Gaining consent after the individual has had experience with a particular service or research study may allow subjects to make better-informed decisions than a sweeping form of secured consent. Furthermore, sustained consent can allow participants to make more granular decisions about what they would be willing to share, with a better understanding of the context, rather than a single consent form or EULA being considered *carte blanche* for unconstrained data collection.

The distinction between secured and sustained consent reveals a tension between two variables: *burden*—the time spent and cognitive load required to negotiate the consent process, and *accuracy*—the extent to which the effect of a consent decision corresponds with a person's expectations. While a secured instrument such as a consent form minimises the burden on the individual, with only a single form to read and comprehend, the accuracy is impossible to discern, with no process for validating that consent decision in context, nor to assess the individual's comprehension of what they have agreed to. Conversely, while a sustained approach—such as asking someone whether they are willing for each item of personal data to be used for data mining activities—may improve accuracy, the added burden is significant and can be frustrating, contributing to attrition, which is particularly problematic in longitudinal studies [17].

In some domains other than data mining, this distinction has already been applied. In biomedical research, the consent to the use of samples is commonly distinguished as being broad or dynamic. Broad consent allows samples to be used for a range of experiments within an agreed framework without consent being explicitly required [50], whereas dynamic consent involves ongoing engagement with participants, allowing them to see how their samples are used, and permitting renegotiation of consent if the samples are to be used for different studies, or if the participant's wishes change [25]. Despite the differences from the data mining domain, the same consent challenges resonate.

Various researchers have proposed ways of minimising the burden of consent, while simultaneously collecting meaningful and accurate information from people.

Williams et al. look at sharing medical data, enhancing agency with a dynamic consent model that enables control of data electronically, and improved legibility by providing patients with information about how their data are used [54]. Gomer et al. propose the use of agents who make consent decisions on behalf of individuals to reduce the burden placed on them, based on preferences they have expressed which are periodically reviewed [13]. Moran et al. suggest that consent can be negotiated in multi-agent environments by identifying interaction patterns to determine appropriate times to acquire consent [32].

We have discussed legibility as an important aspect of HDI, and Morrison et al. study how to visualise collected data to research participants [33]. Personalised visualisations led participants in an empirical study to exit the study earlier, which might mean that secured consent was leading participants to continue beyond the appropriate level of data collection. In much earlier work, Patrick looked at presenting user agreements contextually (rather than at the beginning of a transaction as in secured consent) and developed a web-based widget to do so [40].

Such dynamic approaches to consent are not universally supported. Steinsbekk et al. suggest that where data are re-used for multiple studies, there is no need to acquire consent for each one where there are no significant ethical departures because of the extra burden, arguing that it puts greater responsibility on individuals to discern whether a study is ethically appropriate than existing governance structures [50].

In previous work, we have developed a method for acquiring consent which aims to maximise accuracy and minimise burden, satisfying both requirements, bringing some of the principles of dynamic consent to the data mining domain [22], while aiming to maintain contextual integrity by respecting prevailing social norms. While many of the consent approaches discussed in this chapter may satisfy a legal requirement, it is not clear that this satisfies the expectations of individuals or society at large, and thus may violate contextual integrity.

In a user study, we examine whether prevailing norms representing willingness to share specific types of Facebook data with researchers, along with limited data about an individual's consent preferences, can be used to minimise burden and maximise accuracy. The performance of these measures were compared to two controls: one for accuracy and for burden. In the first instance, a sustained consent process which gains permission for the use of each individual item of Facebook data would maximise accuracy while pushing great burden on to the individual. Secondly, a single consent checkbox minimises burden, while also potentially minimising the accuracy of the method. The contextual integrity method works similarly to this approach, by asking a series of consent questions until the individual's conformity to the social norm can be inferred, at which point no more questions are asked.

For 27.7% of participants, this method is able to achieve high accuracy of 96.5% while reducing their burden by an average of 41.1%. This is highlighted in Fig. 1, showing a cluster of norm-conformant participants achieving high accuracy and low burden. This indicates that for this segment of the population, the contextual integrity approach both improves accuracy and reduces burden compared to their respective controls. While this does indicate the approach is not suitable for all peo-

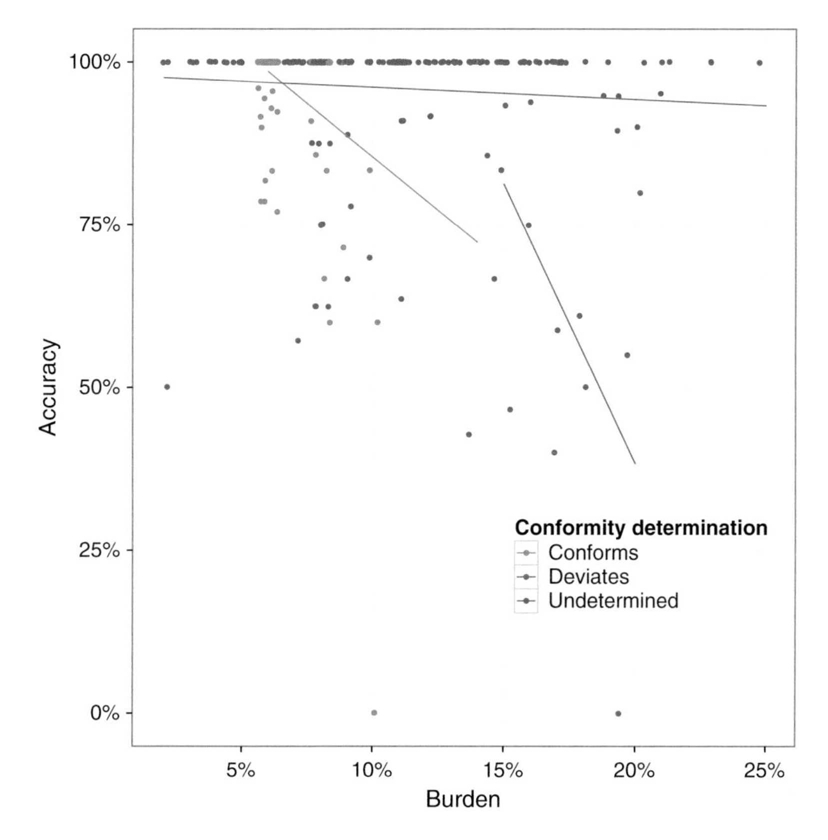

Fig. 1 Scatterplot showing the relationship between norm conformity and accuracy. As indicated by the cluster after five questions (5% burden), high accuracy can be preserved when norm conformity is detected quickly, although the technique is not useful for people who are highly norm deviant. Note that the points have been jittered to improve readability [22]

ple, norm conformity is able to be quickly determined within six questions. Where one does not conform to the norm, the sustained approach can be automatically used as a fallback, which maintains accuracy at the cost of a greater time burden on the individual. Even in less optimal cases, the technique can reduce the burden by an average of 21.9%.

While the technique assessed in this user study is prototypical in its nature, it highlights the potential value of examining alternative means of acquiring consent, which has seen little innovation in both academic and commercial domains. Moreover, while this technique is not universally applicable, this only highlights that the diversity of perspectives, willingness to engage, and ability to comprehend consent language requires a plurality of approaches.

5 Discussion

In this chapter we have illustrated how data mining activities, in both academic and commercial contexts, are often opaque by design. Insufficient consent mechanisms can prevent people from understanding what they are agreeing to, particularly where the scope of the data collected or with whom it is shared is changed without consent being renegotiated. Indeed, as in our three case studies, consent is often not sought at all.

We have considered the impacts of opaque data mining in terms of legibility, agency, and negotiability. We now propose some best practices for conducting data mining which aim to satisfy these three themes.

5.1 *Legibility*

In order to make data mining more acceptable, it is not sufficient to simply make processes more transparent. Revealing the algorithms, signals, and inferences may satisfy a particularly technically competent audience, but for most people does not help them understand what happens to their data, in order to make an informed decision over whether to consent, or how they can act with any agency.

The incoming General Data Protection Regulation (GDPR) in the European Union requires consent language to be concise, transparent, intelligible and easily accessible [43], which as indicated in the literature, is currently not a universal practice. As highlighted in our three case studies, the absence of any meaningful consent enabling data to be used beyond its original context, such as a hospital or social network site, is unacceptable. Even without adopting more sophisticated approaches to consent as discussed in Sect. 4, techniques to notify and reacquire consent such that people are aware and engaged with ongoing data mining practices can be deployed. As discussed earlier, a practical first step is to ensure all consent documents can be understood by a broad spectrum of the population.

5.2 *Agency*

Assuming that legibility has been satisfied, and people are able to understand how their data are being used, the next challenge is to ensure people are able to act autonomously and control how their data are used beyond a single consent decision. Some ways of enabling this include ensuring people can subsequently revoke their consent for their data to be used at any time, without necessarily being precipitated by a change in how the data are used. In the GDPR, this is enshrined through the right to be forgotten [43] that includes the cascading revocation of data between data controllers.

Legibility can also enable agency by allowing people to act in a certain way in order to selectively allow particular inferences to be made. By being able to choose what they are willing to share with a data collector in order to satisfy their own utility, some of the power balance can be restored, which has been previously tipped towards the data collector who is able to conduct analyses at a scale beyond any individual subject's capabilities.

5.3 Negotiability

As discussed in Sect. 4, Nissenbaum's contextual integrity [38] can be used to detect privacy violations when the terms of data-handling have changed in such a way that existing norms are breached. The principle of negotiability is key to preventing this, by allowing people to make ongoing decisions about how their data are used as contexts evolve, whether their own, environmentally, or that of the data collector.

Dynamic consent in the biobanks context [25] could be adapted to allow data subjects to be notified and review how their data are being used, whether for new purposes or shared with new actors, allowing consent to be renegotiated. Our consent method informed by contextual integrity [22] is one such approach which aims to tackle this problem, by allowing people to make granular consent decisions without being overwhelmed. Adopting the principles of the GDPR, which emphasises dynamic consent, can support negotiability, with guidance made available for organisations wishing to apply these principles [51].

6 Conclusion

Data mining is an increasingly pervasive part of daily life, with the large-scale collection, processing, and distribution of personal data being used for myriad purposes. In this chapter, we have outlined how this often happens without consent, or the consent instruments used are overly complex or inappropriate. Data mining is outgrowing existing regulatory and ethical governance structures, and risks violating entrenched norms about the acceptable use of personal data, as illustrated in case studies spanning the commercial and academic spheres. We argue that organisations involved in data mining should provide legible consent information such that people can understand what they are agreeing to, support people's agency by allowing them to selectively consent to different processing activities, and to support negotiability by allowing people to review or revoke their consent as the context of the data mining changes. We have discussed recent work which dynamically negotiates consent, including a technique which leverages social norms to acquire granular consent without overburdening people. We call for greater public debate to negotiate these new social norms collectively, rather than allowing organisations to unilaterally impose new practices without oversight.

Acknowledgements This work was supported by the Engineering and Physical Sciences Research Council [grant number EP/L021285/1].

References

1. Akkad, A., Jackson, C., Kenyon, S., Dixon-Woods, M., Taub, N., Habiba, M.: Patients' perceptions of written consent: questionnaire study. Br. Med. J. **333**(7567), 528+ (2006). doi:10.1136/bmj.38922.516204.55
2. Ayalon, O., Toch, E.: Retrospective privacy: managing longitudinal privacy in online social networks. In: Proceedings of the Ninth Symposium on Usable Privacy and Security. ACM, New York (2013). doi:10.1145/2501604.2501608
3. Barnes, S.B.: A privacy paradox: social networking in the United States. First Monday **11**(9) (2006). doi:10.5210/fm.v11i9.1394
4. Bauer, L., Cranor, L.F., Komanduri, S., Mazurek, M.L., Reiter, M.K., Sleeper, M., Ur, B.: The post anachronism: the temporal dimension of Facebook privacy. In: Proceedings of the 12th ACM Workshop on Workshop on Privacy in the Electronic Society, pp. 1–12. ACM, New York (2013). doi:10.1145/2517840.2517859
5. Berg, J.W., Appelbaum, P.S.: Informed Consent Legal Theory and Clinical Practice. Oxford University Press, Oxford (2001)
6. Brown, I., Brown, L., Korff, D.: Using NHS patient data for research without consent. Law Innov. Technol. **2**(2), 219–258 (2010). doi:10.5235/175799610794046186
7. Carmichael, L., Stalla-Bourdillon, S., Staab, S.: Data mining and automated discrimination: a mixed legal/technical perspective. IEEE Intell. Syst. **31**(6), 51–55 (2016). doi:10.1109/mis.2016.96
8. Donovan-Kicken, E., Mackert, M., Guinn, T.D., Tollison, A.C., Breckinridge, B.: Sources of patient uncertainty when reviewing medical disclosure and consent documentation. Patient Educ. Couns. **90**(2), 254–260 (2013). doi:10.1016/j.pec.2012.10.007
9. Eslami, M., Rickman, A., Vaccaro, K., Aleyasen, A., Vuong, A., Karahalios, K., Hamilton, K., Sandvig, C.: "I always assumed that I wasn't really that close to [her]": reasoning about invisible algorithms in news feeds. In: Proceedings of the 33rd Annual ACM Conference on Human Factors in Computing Systems, CHI '15, pp. 153–162. ACM, New York (2015). doi:10.1145/2702123.2702556
10. European Parliament and the Council of the European Union: Directive 95/46/EC of the European Parliament and of the Council of 24 October 1995 on the protection of individuals with regard to the processing of personal data and on the free movement of such data. Off. J. Eur. Union **L 281**, 0031–0050 (1995)
11. Fayyad, U., Piatetsky-Shapiro, G., Smyth, P.: From data mining to knowledge discovery in databases. AI Mag. **17**(3) (1996). doi:10.1609/aimag.v17i3.1230
12. Friedman, B., Lin, P., Miller, J.K.: Informed consent by design. In: Cranor, L.F., Garfinkel, S. (eds.) Security and Usability, Chap. 24, pp. 495–521. O'Reilly Media, Sebastopol (2005)
13. Gomer, R., Schraefel, M.C., Gerding, E.: Consenting agents: semi-autonomous interactions for ubiquitous consent. In: Proceedings of the 2014 ACM International Joint Conference on Pervasive and Ubiquitous Computing: Adjunct Publication, UbiComp '14 Adjunct, pp. 653–658. ACM, New York (2014). doi:10.1145/2638728.2641682
14. Hamnes, B., van Eijk-Hustings, Y., Primdahl, J.: Readability of patient information and consent documents in rheumatological studies. BMC Med. Ethics **17**(1) (2016). doi:10.1186/s12910-016-0126-0
15. Hastie, T., Tibshirani, R., Friedman, J.: The Elements of Statistical Learning: Data Mining, Inference, and Prediction, corrected edn. Springer, New York (2003)

16. Heimbach, I., Gottschlich, J., Hinz, O.: The value of user's Facebook profile data for product recommendation generation. Electr. Mark. **25**(2), 125–138 (2015). doi:10.1007/s12525-015-0187-9
17. Hektner, J.M., Schmidt, J.A., Csikszentmihalyi, M.: Experience Sampling Method: Measuring the Quality of Everyday Life. SAGE Publications, Thousand Oaks (2007)
18. Hill, K.: Facebook Added 'Research' To User Agreement 4 Months After Emotion Manipulation Study. http://onforb.es/15DKfGt (2014). Accessed 30 Nov 2016
19. Hoadley, C.M., Xu, H., Lee, J.J., Rosson, M.B.: Privacy as information access and illusory control: the case of the Facebook News Feed privacy outcry. Electron. Commer. Res. Appl. **9**(1), 50–60 (2010). doi:10.1016/j.elerap.2009.05.001
20. Hodson, H.: Did Google's NHS patient data deal need ethical approval?. https://www.newscientist.com/article/2088056-did-googles-nhs-patient-data-deal-need-ethical-approval/ (2016). Accessed 30 Nov 2016
21. Hodson, H.: Google knows your ills. New Sci. **230**(3072), 22–23 (2016). doi:10.1016/s0262-4079(16)30809-0
22. Hutton, L., Henderson, T.: "I didn't sign up for this!": informed consent in social network research. In: Proceedings of the 9th International AAAI Conference on Web and Social Media, pp. 178–187 (2015). http://www.aaai.org/ocs/index.php/ICWSM/ICWSM15/paper/view/10493
23. Jackman, M., Kanerva, L.: Evolving the IRB: building robust review for industry research. Wash. Lee Law Rev. Online **72**(3), 442–457 (2016). http://scholarlycommons.law.wlu.edu/wlulr-online/vol72/iss3/8/
24. Kang, J., Shilton, K., Estrin, D., Burke, J., Hansen, M.: Self-surveillance privacy. Iowa Law Rev. **97**(3), 809–848 (2012). doi:10.2139/ssrn.1729332
25. Kaye, J., Whitley, E.A., Lund, D., Morrison, M., Teare, H., Melham, K.: Dynamic consent: a patient interface for twenty-first century research networks. Eur. J. Hum. Genet. **23**(2), 141–146 (2014). doi:10.1038/ejhg.2014.71
26. Kramer, A.D.I., Guillory, J.E., Hancock, J.T.: Experimental evidence of massive-scale emotional contagion through social networks. Proc. Natl. Acad. Sci. **111**(24), 8788–8790 (2014). doi:10.1073/pnas.1320040111
27. Lewis, K., Kaufman, J., Gonzalez, M., Wimmer, A., Christakis, N.: Tastes, ties, and time: a new social network dataset using Facebook.com. Soc. Netw. **30**(4), 330–342 (2008). doi:10.1016/j.socnet.2008.07.002
28. Luger, E., Rodden, T.: An informed view on consent for UbiComp. In: Proceedings of the 2013 ACM International Joint Conference on Pervasive and Ubiquitous Computing, pp. 529–538. ACM, New York (2013). doi:10.1145/2493432.2493446
29. Luger, E., Moran, S., Rodden, T.: Consent for all: revealing the hidden complexity of terms and conditions. In: Proceedings of the SIGCHI Conference on Human Factors in Computing Systems, pp. 2687–2696. ACM, New York (2013). doi:10.1145/2470654.2481371
30. McDonald, A.M., Cranor, L.F.: The cost of reading privacy policies. I/S: J. Law Policy Inf. Soc. **4**(3), 540–565 (2008). http://www.is-journal.org/files/2012/02/Cranor_Formatted_Final.pdf
31. Miller, F.G., Wertheimer, A.: Preface to a theory of consent transactions: beyond valid consent. In: Miller, F., Wertheimer, A. (eds.) The Ethics of Consent, Chap. 4, pp. 79–105. Oxford University Press, Oxford (2009). doi:10.1093/acprof:oso/9780195335149.003.0004
32. Moran, S., Luger, E., Rodden, T.: Exploring patterns as a framework for embedding consent mechanisms in human-agent collectives. In: Ślęzak, D., Schaefer, G., Vuong, S., Kim, Y.S. (eds.) Active Media Technology. Lecture Notes in Computer Science, vol. 8610, pp. 475–486. Springer International Publishing, New York (2014). doi:10.1007/978-3-319-09912-5_40
33. Morrison, A., McMillan, D., Chalmers, M.: Improving consent in large scale mobile HCI through personalised representations of data. In: Proceedings of the 8th Nordic Conference on Human-Computer Interaction: Fun, Fast, Foundational, NordiCHI '14, pp. 471–480. ACM, New York (2014). doi:10.1145/2639189.2639239

34. Mortier, R., Haddadi, H., Henderson, T., McAuley, D., Crowcroft, J., Crabtree, A.: Human-data interaction. In: Soegaard, M., Dam, R.F. (eds.) Encyclopedia of Human-Computer Interaction, Chap. 41. Interaction Design Foundation, Aarhus (2016). https://www.interaction-design.org/literature/book/the-encyclopedia-of-human-computer-interaction-2nd-ed/human-data-interaction
35. Munteanu, C., Molyneaux, H., Moncur, W., Romero, M., O'Donnell, S., Vines, J.: Situational ethics: Re-thinking approaches to formal ethics requirements for human-computer interaction. In: Proceedings of the 33rd Annual ACM Conference on Human Factors in Computing Systems, pp. 105–114. ACM, New York (2015). doi:10.1145/2702123.2702481
36. Napoli, P.M.: Social media and the public interest: governance of news platforms in the realm of individual and algorithmic gatekeepers. Telecommun. Policy **39**(9), 751–760 (2015). doi:10.1016/j.telpol.2014.12.003
37. Narayanan, A., Shmatikov, V.: De-anonymizing social networks. In: Proceedings of the IEEE Symposium on Security and Privacy, pp. 173–187. IEEE, Los Alamitos, CA (2009). doi:10.1109/sp.2009.22
38. Nissenbaum, H.: Privacy in Context: Technology, Policy, and the Integrity of Social Life. Stanford Law Books, Stanford, CA (2009)
39. Obar, J.A., Oeldorf-Hirsch, A.: The biggest lie on the internet: ignoring the privacy policies and terms of service policies of social networking services. Social Science Research Network Working Paper Series (2016). doi:10.2139/ssrn.2757465
40. Patrick, A.: Just-in-time click-through agreements: interface widgets for confirming informed, unambiguous consent. J. Internet Law **9**(3), 17–19 (2005). http://nparc.cisti-icist.nrc-cnrc.gc.ca/npsi/ctrl?action=rtdoc&an=8914195&lang=en
41. Pitofsky, R., Anthony, S.F., Thompson, M.W., Swindle, O., Leary, T.B.: Privacy online: fair information practices in the electronic marketplace: a report to congress. Security. http://www.ftc.gov/reports/privacy2000/privacy2000.pdf (2000)
42. Recuber, T.: From obedience to contagion: discourses of power in Milgram, Zimbardo, and the Facebook experiment. Res. Ethics **12**(1), 44–54 (2016). doi:10.1177/1747016115579533
43. Regulation (EU) 2016/679 of the European Parliament and of the Council of 27 April 2016 on the protection of natural persons with regard to the processing of personal data and on the free movement of such data, and repealing Directive 95/46/EC (General Data Protection Regulation), pp. 1–88. Off. J. Eur. Union **L119/1** (2016)
44. Sankar, P., Mora, S., Merz, J.F., Jones, N.L.: Patient perspectives of medical confidentiality. J. Gen. Inter. Med. **18**(8), 659–669 (2003). doi:10.1046/j.1525-1497.2003.20823.x
45. Selinger, E., Hartzog, W.: Facebook's emotional contagion study and the ethical problem of co-opted identity in mediated environments where users lack control. Res. Ethics **12**(1), 35–43 (2016). doi:10.1177/1747016115579531
46. Sleeper, M., Balebako, R., Das, S., McConahy, A.L., Wiese, J., Cranor, L.F.: The post that wasn't: exploring self-censorship on Facebook. In: Proceedings of the 2013 Conference on Computer Supported Cooperative Work, CSCW 2013, pp. 793–802. ACM, New York (2013). doi:10.1145/2441776.2441865
47. Solove, D.J.: Privacy self-management and the consent dilemma. Harv. Law Rev. **126**(7), 1880–1903 (2013). http://heinonline.org/HOL/Page?handle=hein.journals/hlr126&id=&page=&collection=journals&id=1910
48. Staiano, J., Oliver, N., Lepri, B., de Oliveira, R., Caraviello, M., Sebe, N.: Money walks: a human-centric study on the economics of personal mobile data. In: Proceedings of Ubicomp 2014 (2014). doi:10.1145/2632048.2632074
49. Steinke, G.: Data privacy approaches from US and EU perspectives. Telematics Inform. **19**(2), 193–200 (2002). doi:10.1016/s0736-5853(01)00013-2
50. Steinsbekk, K.S., Kare Myskja, B., Solberg, B.: Broad consent versus dynamic consent in biobank research: is passive participation an ethical problem? Eur. J. Hum. Genet. **21**(9), 897–902 (2013). doi:10.1038/ejhg.2012.282
51. Tankard, C.: What the GDPR means for businesses. Netw. Secur. **2016**(6), 5–8 (2016). doi:10.1016/s1353-4858(16)30056-3

52. Vitak, J., Shilton, K., Ashktorab, Z.: Beyond the Belmont Principles: ethical challenges, practices, and beliefs in the online data research community. In: Proceedings of the 19th ACM Conference on Computer-Supported Cooperative Work and Social Computing, pp. 941–953. ACM, New York (2016). doi:10.1145/2818048.2820078
53. Vučemilo, L., Borovečki, A.: Readability and content assessment of informed consent forms for medical procedures in Croatia. PLoS One **10**(9), e0138,017+ (2015). doi:10.1371/journal.pone.0138017
54. Williams, H., Spencer, K., Sanders, C., Lund, D., Whitley, E.A., Kaye, J., Dixon, W.G.: Dynamic consent: a possible solution to improve patient confidence and trust in how electronic patient records are used in medical research. JMIR Med. Inform. **3**(1), e3+ (2015). doi:10.2196/medinform.3525
55. World Economic Forum: Personal data: the emergence of a new asset class. http://www.weforum.org/reports/personal-data-emergence-new-asset-class (2011)
56. Zimmer, M.: "But the data is already public": on the ethics of research in Facebook. Ethics Inf. Technol. **12**(4), 313–325 (2010). doi:10.1007/s10676-010-9227-5

Regulating Algorithms' Regulation? First Ethico-Legal Principles, Problems, and Opportunities of Algorithms

Giovanni Comandè

Abstract Algorithms are regularly used for mining data, offering unexplored patterns, and deep non-causal analyses in what we term the "classifying society". In the classifying society individuals are no longer targetable as individuals but are instead selectively addressed for the way in which some clusters of data that they (one or more of their devices) share with a given model fit in to the analytical model itself. This way the classifying society might bypass data protection as we know it. Thus, we argue for a change of paradigm: to consider and regulate anonymities—not only identities—in data protection. This requires a combined regulatory approach that blends together (1) the reinterpretation of existing legal rules in light of the central role of privacy in the classifying society; (2) the promotion of disruptive technologies for disruptive new business models enabling more market control by data subjects over their own data; and, eventually, (3) new rules aiming, among other things, to provide to data generated by individuals some form of property protection similar to that enjoyed by the generation of data and models by businesses (e.g. trade secrets). The blend would be completed by (4) the timely insertion of ethical principles in the very generation of the algorithms sustaining the classifying society.

Abbreviations

AI	Artificial intelligence
CAL. BUS & PROF. CODE	California business and professions code
CAL. CIV. CODE	California civil code
CONN. GEN. STAT. ANN.	Connecticut general statutes annotated
DAS	Domain awareness system
DHS	U.S. Department of Homeland Security
DNA	Deoxyribonucleic acid
EDPS	European data protection supervisor
EFF	Electronic Frontier Foundation

G. Comandè (✉)
Scuola Superiore Sant'Anna Pisa, Piazza Martiri della Libertà 33, 56127, Pisa, Italy
e-mail: g.comande@santannapisa.it

T. Cerquitelli et al. (eds.), *Transparent Data Mining for Big and Small Data*,
Studies in Big Data 32, DOI 10.1007/978-3-319-54024-5_8

EU	European Union
EU GDPR	European Union general data protection regulation
EUCJ	European Union Court of Justice
FTC	Federal Trade Commission
GA. CODE ANN.	Code of Georgia annotated
GPS	Global positioning system
GSM	Global system for mobile communications
GSMA	GSM Association
ICT	Information and communications technology
NSA	National Security Agency
PETs	Privacy-enhancing technologies
PPTCs	Privacy policy terms and conditions
SDNY	United States District Court for the Southern District of New York
ToS	Terms of service
WEF	World Economic Forum
WPF	World Privacy Forum

> "Today's trends have opened an entirely new chapter, and there is a need to explore whether the principles are robust enough for the digital age. The notion of personal data itself is likely to change radically as technology increasingly allows individuals to be re-identified from supposedly anonymous data. In addition, machine learning and the merging of human and artificial intelligence will undermine concepts of the individual's rights and responsibility" [1, p. 13].

1 Introduction

Today's technologies enable unprecedented exploitation of information, be it small or big data, for any thinkable purpose, but mostly in business [2, 3] and surveillance [4] with the ensuing juridical and ethical anxieties.

Algorithms are regularly used for mining data, offering unexplored patterns and deep non-causal analyses to those businesses able to exploit these advances. They are advocated as advanced tools for regulation and legal problem-solving with innovative evidence gathering and analytical capacities.

Yet, these innovations need to be properly framed in the existing legal background, fit in to the existing set of constitutional guarantees of fundamental rights and freedoms, and coherently related to existing policies in order to enable our societies to reap the richness of big and open data while equally empowering all players.

Indeed, if the past is normally the prologue of the future, when our everyday life is possibly influenced by filtering bubbles and un-verifiable analytics (in a word: it is heteronomously "pre-set"), a clear ethical and legal response is desperately needed to govern the phenomenon. Without it, our societies will waver between two extremes—either ignoring the problem or fearing it unduly. Without a clear transnational ethical and legal framework, we risk either losing the immense possibilities entailed by big and small data analytics or surrendering the very gist of our praised rights and freedoms.

To secure the benefits of these innovations and avoid the natural perils every innovation brings along, this paper makes a call for regulating algorithms and expanding their use in legal problem-solving at various levels by first exploring existing legal rules. Accordingly, this paper, building on existing literature: (1) frames the main legal and ethical issues raised by the increasing use of algorithms in information society, in digital economy, and in heightened public and private surveillance; (2) sets the basic principles that must govern them globally in the mentioned domains; (3) calls for exploring additional uses of algorithms' evidence-based legal problem-solving facility in order to expand their deployment on behalf of the public interest (e.g. by serving pharmacovigilance) and to empower individuals in governing their data production and their data flow.

It is important to stress that the tremendous benefits of big data—for instance, of the interplay among data mining and machine learning—are not questioned here. To the contrary, this paper asserts that in order to reap all the promised richness of these benefits it is important to correctly frame the reality of the interplay of these technologies. It is also important to emphasize that data protection, whatever form it takes in a given legal system, can be a key to development, rather than an obstacle.[1] There is no possible alternative to sustaining such a bold statement, since in the digital age static profiles are basically disappearing due to the expansion of machine learning and artificial intelligence. Among other things, such an evolution implies that the classification model used in any given moment no longer exists, as such, in a (relatively contemporaneous) subsequent one. Accountability, then, requires both the technical (already extant) and the legal ability to establish with certainty in each moment which profile has been used in the decision process.

Thus, the emerging regulatory approach must necessarily blend together various legal, technological, and economic strategies for which the time frame is of crucial importance.

The EDPS [1, p. 13] rightly stressed that "The EU in particular now has a 'critical window' before mass adoption of these technologies to build the values into digital structures which will define our society. This requires a new assessment of whether the potential benefits of the new technologies really depend on the collection and analysis of the personally-identifiable information of billions of individuals. Such an

[1] "Contrary to some claims, privacy and data protection are a platform for a sustainable and dynamic digital environment, not an obstacle" [1, p. 9].

assessment could challenge developers to design products which depersonalise in real time huge volumes of unorganized information making it harder or impossible to single out an individual." Yet, at this stage we should wonder whether it is more a matter of privacy protection or dignity threatened by the *biased targeting of anonymities*, not just individuals.

2 The Classifying Society

The Free Dictionary defines the verb "to classify" as: "to arrange or organize according to class or category." Google, faithful to its nature, clarifies: "[to] arrange (a group of people or things) in classes or categories according to shared qualities or characteristics".

According to the Cambridge Advanced Learner's Dictionary, the verb "classify" originally referred, however, to "things", not to humans.[2] The societies in which we live have altered the original meaning, extending and mostly referring it to humans. Such a phenomenon is apparent well beyond big data[3] and the use of analytics. For instance, in medicine and research the term "classification" now applies to old and new (e.g. emerging from DNA sequencing) shared qualities and characteristics that are compiled with the aim of enabling increasingly personalized advertising, medicine, drugs, and treatments.[4]

The rising possibilities offered by machine learning, data storage, and computing ability are entirely changing both our real and virtual landscapes. Living and being treated according to one or more class is the gist of our daily life as personalized advertising[5] and the attempt by companies to anticipate our desires and needs[6] clearly illustrate.

Often every group of "shared qualities or characteristics" even takes priority over our actual personal identity. For instance, due to a group of collected characteristics,

---

[2]Cambridge Advanced Learner's Dictionary & Thesaurus of the Cambridge University Press specifies as the first meaning: "to divide things into groups according to their type: The books in the library are classified by/according to subject. Biologists classify animals and plants into different groups".

[3]On the risks and benefits of big data see e.g., Tene and Polonetsky [5]; *contra* Ohm [6].

[4]"Personalization is using more (demographic, but also behavioral) information about a particular individual to tailor predictions to that individual. Examples are Google's search results based on individual's cookies or GMail contents" [7, p. 261]. See also https://www.google.com/experimental/gmailfieldtrial.

[5]"Algorithms nowadays define how we are seen, by providing a digital lens, tailored by statistics and other biases." [7, p. 256].

[6]Amazon for instance is aiming at shipping goods to us even before we place an order [8]. This approach is very similar to Google attempting to understand what we want before we know we want it. "Google is a system of almost universal surveillance, yet it operates so quietly that at times it's hard to discern" [9, p. 84].

we can be classified and targeted accordingly as (potential) terrorists, poor lenders,[7] breast cancer patients, candidates for a specific drug/product, or potentially pregnant women.[8] These classifications can be produced and used without us having the faintest clue of their existence, even though we are not actual terrorists (perhaps, for instance, we like to travel to particular areas and happen to have an Arab-sounding name), we are affluent (but prefer to live in a much poorer neighbourhood), we do not have breast cancer, we have no need for or interest whatsoever in a drug/product, we are not even female, or we are not a sex addict (see infra).

The examples referred to above are well documented in literature and have been chosen to illustrate that classifications characterize every corner of our daily life.[9] They expose most of the legal problems reported by scholars and concerned institutions. Most of these problems revolve around notions of privacy, surveillance, and danger to freedom of speech,[10] ...

Yet, literature to date has failed to discuss the very fact that the classifying society we live in threatens to make our actual identities irrelevant, fulfilling an old prophecy in a cartoon from a leading magazine. This cartoon displayed a dog facing a computer while declaring: "on internet nobody cares you are actually a dog". With hindsight, we could now add "if you are classified as a dog it is irrelevant you are the human owner of a specific kind of dog". Indeed, Mac users are advertised higher prices regardless of whether they are personally identified as affluent [18]. Although in some countries (such as the USA) price discrimination and customer-steering are not forbidden unless they involve prohibited forms of discrimination,[11] we should begin to question the ethics of such processes once they are fully automatic and

[7]See for more examples Citron and Pasquale [10].

[8]By using previous direct interaction, Target knew a teenage girl was pregnant well before her family did [11].

[9]See, for instance, the following list of horrors in Gray and Citron [12, p. 81, footnotes omitted]: "Employers have refused to interview or hire individuals based on incorrect or misleading personal information obtained through surveillance technologies. Governmental data-mining systems have flagged innocent individuals as persons of interest, leading to their erroneous classifications as terrorists or security threats. ... In one case, Maryland state police exploited their access to fusion centers in order to conduct surveillance of human rights groups, peace activists, and death penalty opponents over a 19 month period. Fifty-three political activists eventually were classified as 'terrorists,' including two Catholic nuns and a Democratic candidate for local office. The fusion center subsequently shared these erroneous terrorist classifications with federal drug enforcement, law enforcement databases, and the National Security Administration, all without affording the innocent targets any opportunity to know, much less correct, the record."

[10]On the chilling effect of dataveillance for autonomy and freedom of speech see, for instance, in literature [13–17].

[11]The limits of antidiscrimination law to cope with data-driven discrimination have been already highlighted by Barocas and Selbst [19].

unknown to the target.[12] In addition, they lock individuals into myriad anonymous models based upon traits that they might not be able to change, even if, theoretically, such traits are not as fixed as a fingerprint—as, for instance, where the use of an Apple product is needed for the specific kind of work performed, or happens to be required by ones' employer.

Even in those instances in which the disparate classifying impact is technically legal it can be ethically questionable, to say the least. Such questionability does not depend so much on the alleged aura of infallibility attributed to automatic computerized decision-making [23, p. 675] as on its pervasiveness and unavoidability.

Apparently there is nothing new under the sun. Humans, for good or bad, have always been classified. Classes have been a mode of government as well. What is different about the classificatory price discrimination (and not only price!)[13] that can be systematically produced today—different from the old merchant assessing the worth of my dresses adjusted according to the price requested—is discrimination's dimension, pace, pervasiveness and relative economy.[14]

Literature has also failed to address another key issue: due to the role of big data and the use of algorithms, actual personal identification as a requirement for targeting individuals is becoming irrelevant in the classifying society. The emerging targets are clusters of data (the matching combination between a group of data and a given model), not physical individuals. The models are designed to identify clusters of characteristics, not the actual individuals who possess them—and even when the model incorrectly identifies a combination, it is refining its own code,[15] perfecting its ability to match clusters of shared characteristics in an infinite loop.

Classification based on "our" "shared characteristics" covers every corner of us (what we are and what we do, let alone how we do it) once we can be targeted at no cost as being left-handed, for instance. Yet, the power (resources, technology, and network economies) to do such classifying is quickly becoming concentrated in fewer hands than ever.[16] The pace of classification is so rapid as to have begun happening in real time and at virtually no cost for some players. Its pervasiveness is evidenced by the impact our web searches have on our personalized

[12]Some forms of notification at least have already been advocated in the context of the debate surrounding the USA's Fourth Amendment [20]. In the EU, specific rules on automation are in place [21]. However, some authors claim that automation as such does not require higher scrutiny [22].

[13]See also Moss [24], stressing the ability of algorithms to discriminate "in practically and legally analogous ways to a real world real estate agent".

[14]"It is not just the amount of data but also novel ways to analyze this data that change the playing field of any single individual in the information battle against big companies and governments. Data is becoming a key element for profit and control, and computers gain in authority" [7, p. 256; 25].

[15]See infra footnotes 73–85 and accompanying text.

[16]According to EU Competition Commissioner Margrethe Vestager, the EU Commission is considering the proposal of a specific directive on big data.

advertising on the next web site we visit.[17] The progressive switch to digital content and services makes this process even faster and easier: "Evolving integration technologies and processing power have provided organizations the ability to create more sophisticated and in-depth individual profiles based on one's online and offline behaviours" [27, p. 2].[18] Moreover, we are getting increasingly used to the myth of receiving personalized services for free[19] and to alternatives to the surrender of our data not being offered [29].

Classifications are not problematic by definition, but some of their modes of production or uses might be. In particular, this applies extensively to the digital domain, wherein transactions are a continuum linking the parties (businesses and "their" customers)—a reality that is clearly illustrated by the continuous unilateral changes made to Terms of Service (ToS) and Privacy Policy Terms and Conditions (PPTCs) that deeply alter the content of transactions[20] after (apparent) consumption and even contemplate the withdrawal of the product/service without further notice.[21]

On a different account, the expanding possibility of unlocking information by data analysis can have a chilling effect on daily choices when the virtual world meets the real one at a very local level. For instance, a clear representation of the outer extreme of the spectrum introduced at the beginning (fear of technology), could be the hesitation to join discussion groups (on drugs, alcoholism, mental illnesses, sex, and other topics) for fear that such an affiliation can be used in unforeseen ways and somehow disclosed locally (maybe just as a side effect of personalized marketing).[22]

Even when cookies are disabled and the web navigation is "anonymous", traces related to the fingerprints of the devices we use are left. Thus, the elaboration and enrichment of those traces (that are by definition "anonymous") could be related to one or more identifiers of devices, rather than to humans. At this point there is no need to target the owner of the device. It is simpler and less cumbersome from a legal standpoint to target the various evolving clusters of data related to a device or a group of devices instead of the personally identified/able individual. Such a state of affairs calls for further research on the need to "protect" the anonymous (and, to the extent that we are unaware of its existence, imperceptible) identity generated by data analytics.

---

[17] See in general [26].

[18] See also Rajagopal [28].

[19] This is the way in which data collection and sharing is supposedly justified in the eyes of customers.

[20] For a general description, see Perzanowski [30].

[21] Apple [31] for instance imposes the acceptance of the following: "Notwithstanding any other provision of this Agreement, Apple and its licensors reserve the right to change, suspend, remove, or disable access to any products, content, or other materials comprising a part of the Service at any time without notice. In no event will Apple be liable for making these changes. Apple may also impose limits on the use of or access to certain features or portions of the Service, in any case and without notice or liability."

[22] Actually, companies already extensively use algorithms to select employees. For documented cases, see Behm [32].

In this unsettling situation the legal and ethical issues to be tackled could be tentatively subsumed under the related forms of the term "classify".

The adjective "classifiable" and its opposite "non-classifiable" indicate the legal and ethical limits of classification that have so far been made publicly manifest in well-known incidents like the NSA scandal,[23] Facebook tracking,[24] the Domain Awareness System (DAS)[25] and TrapWire software.[26] Yet, the virtual number of classifications and biases is actually infinite [38].

The verb "misclassify" and its related adjectives ("misclassified," "misclassifying") denote instances of error in classification. As paradoxical as it might sound, these algorithms' "mistakes" (false positives and false negatives) reinforce the strength of the classifying society, while simultaneously opening a Pandora's box of new and often unknown/undiscoverable patterns of discrimination.

The verbs "overclassify" and "pre-classify" entail, for instance, risks of excessive and anticipatory classification capable of limiting autonomy, and can certainly be attached to the use of predictive coding in any capacity, be it automatic or human controlled/readable.

Indeed, since literature has clearly illustrated that in the classifying society full personal identification[27] is not necessary to manipulate the environment in which we are operating [29], it is paramount to focus on the tracked/traceable[28] algorithmic "anonymous" identities upon which businesses and governments[29] rely to deal with us—that is, to focus on the classes, those various sets of data that pigeonhole us (these sets of data, are, to use the more appealing technological term, the "models" upon which algorithms act). After all, if a model is already built on data available to our counterpart and only very few of "our" data, even a pseudo-anonymized or fully anonymized model [42] is sufficient to pigeonhole; the classifying society is, across the continents, altogether bypassing data protection as we know it because it is targeting subsets of data fitting specific models rather than individuals expressly. Moreover, as anticipated, these clusters are mostly related to things rather than to individuals. Hence, and for instance, no warning (notice) is needed if the model does not need (therefore it is not even interested) to know it is targeting a given individual in real life; it only needs to identify a cluster of data fitting the model, related to one or more things, regardless of their correspondence with an identified or identifiable

[23]The NSA spying story is nothing new [for the timeline, 33, 34].

[24]FaceBook tracks micro-actions such as mouse movements as well [35].

[25]See Privacy SOS [36].

[26]"TrapWire is a unique, predictive software system designed to detect patterns indicative of terrorist attacks or criminal operations. Utilizing a proprietary, rules-based engine, TrapWire detects, analyzes and alerts on suspicious events as they are collected over periods of time and across multiple locations" [37].

[27]See, on the risk of re-identification of anonymized data, Ohm [39].

[28]"We are constantly tracked by companies and governments; think of smart energy meters, biometric information on passports, number plate tracking, medical record sharing, etc." [7, p. 256].

[29]Often acting synergistically: see Hoofnagle [40]; Singer [41].

individual. Yet, if behind the anonymous subset of data there is an undiscovered real person, that person is effectively being targeted—even if the law as currently formulated does not regard it as such.[30]

These technologies are deployed in the name of a "better user experience," and of fulfilling the future promise of augmented reality and enhanced human ability. Yet, living immersed in the classifying society, we wonder whether the reality that "better matches" our (even anonymous) profiles (shared characteristics in a model) is also a distorted one that reduces our choices.

Such a troubling doubt calls into question the role that law has to play in facing the power of algorithms. It requires exploring both the adapted applications of existing rules and the design of new ones. It also suggests that ethical and legal principles should be shared and embedded in the development of technology [44]. The power to classify in a world that is ruled by classification should entail duties along with rights.

Nevertheless, even the newest regulation on data protection, the EU GDPR,[31] does not address these concerns. To the contrary, it might amplify them, legitimizing the entire classifying society model. But we cannot deal with such issues here.[32]

Finally, the emergence of the classifying society is sustained by an economic argument that is as widespread as it is false: massive data collection rewards companies that provide "free" services and access where an alternate business model (for payment) would fail. First of all, this argument itself makes clear that since the service is rewarded with data collection, it is not actually free at all.[33] Secondly, markets based entirely on a pay-for-service model do exist: the Chinese internet services system is a clear example.

Against this general framework we need to now re-frame a number of the legal issues—already stressed by the literature—generated by the widespread use of algorithms in the classifying society.

2.1 *(Re)sorting Out the Legal Issues*

A large portion of the legal and ethical issues arising from the use of algorithms to read big (or small) data have been already identified by both legal and information

---

[30]See Article 29 Data Protection Working Party [43].

[31]Regulation (EU) 2016/679 of the European Parliament and of the Council of 27 April 2016 on the protection of natural persons with regard to the processing of personal data and on the free movement of such data, and repealing Directive 95/46/EC (General Data Protection Regulation, or "GDPR").

[32]See art. 6 of the GDPR on subsequent processing and pseudo-anonymous data.

[33]The argument that "free" services actually command a price (in data) for their services and the suggestion that "free users should be treated as consumers for the purposes of consumer protection law" has been already advanced [45, pp. 661–662]; on the economic value of data see Aziz and Telang [46].

technology experts. Similarly, there are a variety of taxonomies for algorithms related to their ability to look for hypotheses and explanations, to infer novel knowledge (deductive algorithms) or transform data into models (inductive algorithms), and to use third-party data to colour information (so-called socialized searches such as the ones used by Amazon's book recommendations).[34] All of these taxonomies can be related to the "classify" vocabulary mentioned above, but this is not the task we have here. The following brief overview will explicate the main issues, demonstrating how the two identified deficiencies in the literature call for a different approach and a multilayer regulatory strategy.

A growing literature[35] has already illustrated the need for more transparency in the design and use of data mining, despite the fact that transparency as such can undermine the very goal of predictive practices in a manner that is disruptive to the public interest (for instance by making public the random inspection criteria used to select taxpayers).[36] Nevertheless, what we are questioning here is not the use of classification and algorithms for public goals and by public authorities. Instead, we focus on the pervasiveness of these processes in daily life and at the horizontal level among private entities with strong asymmetries of knowledge and power. Such asymmetries in the use of descriptive and predictive algorithms can generate micro-stigmas which have not been fully explored, let alone uncovered. These micro-stigmas or classifications are so dispersed and undisclosed that mobilization and reaction appear unrealistic. Indeed, and for instance, it is a new stereotype that Apple products users are more affluent than PC users; yet it is a stereotype that escapes existing legal and ethical rules and can lead to higher prices without even triggering data protection rules.

In their reply to Professor Richards, Professors Citron and Gray [54, p. 262] recall the various forms of surveillance leading to "total-information awareness": "coveillance, sousveillance, bureaucratic surveillance, surveillance-industrial complex, panvasive searches, or business intelligence". In stressing the role of fusion centers[37] as a key to this shift to total surveillance,[38] they emphasize the fall of the public/private surveillance divide.

[34]These algorithms use the model built on other people's similar behavioral patterns to make suggestions for us if they think we fit the model (i.e. the classification) they have produced [47].

[35]As beautifully described by Pasquale and Citron [48, p. 1421]: "Unexplained and unchallengeable, Big Data becomes a star chamber... secrecy is a discriminator's best friend: unknown unfairness can never be challenged, let alone corrected". On the importance of transparency and accountability in algorithms of powerful internet intermediaries see also Pasquale [49, 50]. But see, on the role of transparency and the various levels of anonymity, Zarsky [51, 52]; Cohen [53].

[36]The point is clearly illustrated by Zarsky [52].

[37]In their description [54, pp. 264–265, footnotes omitted]: "Fusion centers access specially designed data-broker data-bases containing dossiers on hundreds of millions of individuals, including their Social Security numbers, property records, car rentals, credit reports, postal and shipping records, utility bills, gaming, insurance claims, social network activity, and drug- and food-store records. Some gather biometric data and utilize facial-recognition software."

[38]See the official description [55].

Yet, this is not the Orwellian nightmare of 1984 but a distributed mode of living and "manipulating" in which we are normally targeted indirectly—that is, by matching a subset of our data (that do not need to be personal identifying information in the eyes of law) with the models generated by algorithms during the analysis of large quantities of data (data not necessarily related to us either), all without the need to identify us as individuals.

When we read the actual landscape of the information society in this way, continuous surveillance, as it is strikingly put by Julie E. Cohen [53], alters the way people experience public life.[39] However, the implications for human daily life and activities are deeper and more subversive—and yet mostly unnoticed—in the relationships we develop under private law [29].[40]

Predictive algorithms are applied widely [58][41] and extensively resort to data mining [62, pp. 60–61]. They have intensively and deeply changed the notion of surveillance and have introduced a novel world of covered interaction between individuals and those who sense them (private or public entities or their joint alliance). As Balkin [63, p. 12] puts it, "Government's most important technique of control is no longer watching or threatening to watch. It is analyzing and drawing connections between data.... [D]ata mining technologies allow the state and business enterprises to record perfectly innocent behavior that no one is particularly ashamed of and draw surprisingly powerful inferences about people's behavior, beliefs, and attitudes."[42]

A significant American literature has tackled the problem by examining potential threats to the Fourth Amendment to that country's Constitution [64–67]. Yet, from a global perspective wherein the American constitution does not play a role, it is

[39]See also Cohen [56]. For the psychological impact of surveillance see Karabenick and Knapp [57].

[40]"More unsettling still is the potential combination of surveillance technologies with neuroanalytics to reveal, predict, and manipulate instinctual behavioral patterns of which we are not even aware" [54, p. 265]. Up to the fear that "Based on the technology available, the emergence of a 'Walden 3.0′ with control using positive reinforcements and behavioral engineering seems a natural development." [7, p. 265]. Walden 3.0 would be the manifestation of "Walden Two," the utopian novel written by behavioral psychologist B. F. Skinner (first published in 1948) embracing the proposition that even human behaviour is determined by environmental variables; thus, systematically altering environmental variables can generate a sociocultural system driven by behavioral engineering.

[41]See also Citron [59]; Coleman [60]; Marwick [61, p. 22].

[42]This phenomenon is particularly problematic for jurists since "[o]ne of the great accomplishments of the legal order was holding the sovereign accountable for decisionmaking and giving subjects basic rights, in breakthroughs stretching from Runnymede to the Glorious Revolution of 1688 to the American Revolution. New algorithmic decisionmakers are sovereign over important aspects of individual lives. If law and due process are absent from this field, we are essentially paving the way to a new feudal order of unaccountable reputational intermediaries" [63, p.19].

important to investigate legal and ethical rules that can be applied to private entities as well, beyond their potential involvement in governmental operations.[43]

Despite the fact that scholars have extensively stressed the potential chilling effect[44] of the phenomenon, we claim that the "risk" accompanying the opportunities transcends a disruption of the public arena and casts on individual and collective lives the shadow of an unseen conformation pressure that relentlessly removes the right to be divergent from the available modes of acceptable behaviour, action and reaction, in our commercial and personal dealings. The individual and social impact goes beyond the risks posed by continuous surveillance technology and reduces "the acceptable spectrum of belief and behavior". As it has been stressed, continuous surveillance may result in a "subtle yet fundamental shift in the content of our character". It arrives "not only to chill the expression of eccentric individuality, but also, gradually, to dampen the force of our aspirations to it" [15].[45]

Yet, what we are describing in the classifying society is more than, and different from, surveillance. While surveillance might impair individuals' and groups' ability to "come together to exchange information, share feelings, make plans and act in concert to attain their objectives" [74, p. 125], the classifying society can prevent us from thinking and behaving in our private relationships in a way that diverges from the various models we are pigeonholed into.

Platform neutrality is also in danger in the classifying society since large players and intermediaries have the ability to distort both commerce and the public sphere by leveraging their size or network power or big data.[46] The issue becomes more problematic once we look at the legal protections claimed by intermediaries such as Google.[47] Indeed, the claim of such intermediaries to being merely neutral collectors of preferences[48] is often undermined by the parallel claim they make, as corporations, they enjoy a free speech defence [81] that would allow them to manipulate results in favour of (or contrary to), for example, a political campaign, or a competitor, or a cultural issue. Such a result is somehow acknowledged in the American legal system by Zhang v. Baidu.com Inc.,[49] which affirms that a for-profit platform's selection and arrangement of information is not merely copyrightable, but also represents a form of free speech.[50]

[43] Government actions have triggered and driven a critical debate. See for instance Ramasastry [68]; Slobogin [69]; Solove [70].

[44] On the issue see also Solove [71]; Cate [72]; Strandburg [73].

[45] See also Schwartz [13].

[46] A serious concern shared both in Europe [75] and in the USA [76], stressing the systematic suppression of conservative news.

[47] Of course, it is not only Google that is the problem [77].

[48] It was the case in the Google Anti-defamation league [78]; see also Woan [79]; Wu [80].

[49] 2014 WL 1282730, at 6 (SDNY 2014) ("[A]llowing Plaintiffs to sue Baidu for what are in essence editorial judgments about which political ideas to promote would run afoul of the First Amendment.").

[50] *Contra e.g.*, Case C-131/12.

This latter example of corporate strategic behaviour, switching from one to another self-characterization, illustrates very clearly the need for a transnational legal intervention,[51] or, at least, the need to overcome the stand-alone nature of legal responses that allows a company to use one approach under competition law and another one under constitutional law, for instance.[52]

Yet, this very opportunistic approach should make us wonder if it could be claimed as well by the individuals sensed by algorithms. Since sensed and inferred data are mostly protected as trade or business secrets, it would be worth exploring the possibility of protecting individual anonymities as individual trade secrets in shaping their own bargaining power.

2.2 *On Discrimination, (Dis) Integration, Self-Chilling, and the Need to Protect Anonymities*

Analytical results may have a discriminatory—yet sometimes positive [85]—impact based on unacceptable social factors (such as gender, race, or nationality), on partial information (which being by definition incomplete is typically thereby wrong), on immutable factors over which individuals have no control (genetics, psychological biases, etc.), or on information the impact of which is unknown to the individual (e.g. a specific purchasing pattern). With reference to the latter it can also be the case of a plain and tacit application of a generalization to a specific individual [86, p. 40],[53] or of plain errors in the data mining process [68] amplifying the risks of misclassification or over-classification we have anticipated above.

It is also rather obvious that data mining casts data processing outside of "contextual integrity" [88], magnifying the possibility of a classification based on clusters of data that are taken literally out of context and are therefore misleading. For instance, assume that, for personal reasons unrelated to sexual habits, someone goes through an area where prostitution is exercised. This individual does it at regular times, every week, and at night. The "traffic" in the area forces a slow stop-and-go movement, compatible with attempts to pick up a prostitute, and such a movement is tracked by the geo-localization of one or more of the devices the individual is carrying along. Accordingly, once connected to the web with one of those devices (recognized by one or more of their components' unique identification numbers, such as their Wi-Fi or Bluetooth antenna) the device is targeted by advertising of pornographic materials, dating websites, sexual enhancement drugs, or remedies for sexually transmitted diseases because it fits the model of a sex addict prone to mercenary love.

[51] A need already signalled in the literature [82, 83].

[52] The high risks of enabling such a free speech approach have been highlighted [84; 7, p. 269].

[53] See also Pasquale and Citron [48]; Zarsky [87].

Note that we intentionally switched from referring to the human individual to referring to his/her devices and to the fact data collected, which are directly and exclusively related to the device. There is no need to even relate the cluster of data to personal identifying information such as ownership of the car or contracts for the services provided to the device. Nor is there any need to investigate the gender of the individual since to apply the model it is sufficient that a component in the device is characterized by features "normally" referable to a specific gender (e.g. the pink colour of the smartphone back cover). Of course, the more our society moves towards the Internet of Things, the more the classifications will be related to individual things and groups of related things (related, for instance, because they are regularly physically close to one another—such as a smartphone, a wallet, a laptop, . . .). Targeting groups of related things can be even more accurate than targeting the human individual directly. Indeed, some common sharing of characteristics among things increases the salience of each bit of individual information; for instance their "place of residence" (home) is detected by the fact that they all regularly stay together at the same geo-localized point at night, but variations in their uses offer the possibility of fine tuning the targeting by, for instance, noting that while at "home" the group shows a different behavioural pattern (e.g. related to a different gender/age because a companion or family member is using the device). Accordingly, targeting the device at different hours of the day may expand the reach of the classifying society, again without the need to resort to personal identifying information and with a higher granularity. To the contrary, the classifying society avoids personal identifying information because such information reduces the value of the different clusters of data referable in the various models to the device or to the group of things.

This example clearly illustrates how and why a change of paradigm is required: to consider and regulate anonymities—not only identities—in data protection.

This rather simple example of analytical mismatch also triggers the issue of the harm caused to the undiscovered individual behind the devices. It might be problematic to claim his/her privacy has been intruded. Arguably, the new (erroneous) knowledge produced to sell her/his devices' data for targeted advertising could even be protected as a trade secret. Apparently the model is acting on an error of "perception" not very different from the one a patrolling police officer might have in seeing an individual regularly in a given area, week after week, and deciding to stop him for a "routine control". Yet, while the impact of a regular and lawful search by police is trivial and might end up correcting the misperception, the erroneous fitting to the model of mercenary love user would lead to a variety of problems—such as unease in displaying in public or with relatives, children, spouse, . . . ones' own web searches due to the advertising they trigger and that cannot be corrected by traditional privacy preserving actions (cookies removal, cache cleaning, anonymous web searches,. . .) and reduced ability to exploit searches since "personalized" results pollute the results of such searches on the given device [39].

These are all forms of privacy harm that are difficult to uncover and even more problematic to prove and remedy. In the words of Ryan Calo [89]: "The objective category of privacy harm is the unanticipated or coerced use of information concerning a person against that person. These are negative, external actions justified by reference to personal information. Examples include the unanticipated sale of a user's contact information that results in spam and the leaking of classified information that exposes an undercover intelligence agent." In ordinary people's lives this might involve, for instance, "the government [leverage of] data mining of sensitive personal information to block a citizen from air travel, or when one neighbor forms a negative judgment about another based on gossip" [89, p. 1143].[54]

We claim here that the expanding use of algorithms amplifies the "loss of control over information about oneself or one's own attributes" [89, p. 1134] to a level beyond personal identification. Indeed, in several legal systems data protection is only triggered when personal identifiable information is at stake.[55] In the classifying society vulnerability does not require personal identification.[56]

Data mining is already contributing to a change in the meaning of privacy, expanding the traditional Warren and Brandeis legacy of a "right to be let alone" to privacy as "unwanted sensing" by a third party [90, pp. 225–226]. However, in the classifying society, the interplay among the various techniques for gathering and enriching data for analysis and during the mining process[57] (from self tracking [91] to direct interaction or company logs [11], or the intermediary logging of data such as google searches or cookies or the purchase of data by a broker to integrate and augment existing data bases)[58] has reached a point in which data protection as we know it is powerless and often effectively inapplicable to modern data mining technologies.

Data mining, the "nontrivial process of identifying valid, novel, potentially useful and ultimately understandable patterns in data" [93],[59] is producing a deep change in our thinking paradigm. We used to consider data in contexts, to track them to individuals. Algorithms, big data, and the large computing ability that connect them do not necessarily follow causal patterns and are even able to identify information unknown to the human individual they refer to.[60] Pattern (or "event-based") and subject-based data mining [69, p. 438] search for patterns that describe events and relate them. They result in both descriptive and predictive results. In the latter case, data analysis generates new information based on previous data. This information

[54]According to R. Calo [89] harm must be "*unanticipated* or, if known to the victim, *coerced*".

[55]This is the case for both the EU and the USA. See for instance California Online Privacy Protection Act, CAL. BUS & PROF. CODE §§ 22575–22579 (West 2004) (privacy policy requirement for websites on pages where they collect personally identifiable information); CAL. CIV. CODE §§ 1785.11.2, 1798.29, 1798.82 (West 2009); CONN. GEN. STAT. ANN. § 36a-701b (West 2009 & Supp. 2010); GA. CODE ANN. § 10-1-910, 911 (2009).

[56]See footnotes 35 and 40 and accompanying text.

[57]Mining itself generates new data that change the model and the reading of the clusters.

[58]Clarke [92] defines dataveillance as "the systematic use of personal data systems in the investigation or monitoring of the actions or communications of one or more persons".

[59]However, there are several technical definitions of data mining available but they all refer to the discovery of previously unknown, valid patterns and relationships.

[60]This is the case for a recent study on pancreatic cancer [94].

should be able to predict outcomes (e.g. events or behaviours) by combining previous patterns and new information. The underlying assumption is that results found in older data apply to new ones as well, although no causal explanation is provided even for the first set of results.

Datasets are actively constructed and data do not necessarily come from the same sources, increasing the risks related to de-contextualization. Moreover, since an analyst must define the parameters of the search (and of the dataset to be searched), human biases can be "built in" to the analytical tools (even unwillingly),[61] with the additional effect of their progressive replication and expansion once machine learning is applied. Indeed, if the machine learning process is classified as "non-interpretable" (by humans, sic!), for instance because the machine learned scheme is assessing thousands of variables, there will not be human intervention or a meaningful explanation of why a specific outcome is reached (e.g. a person, object, event, ... is singled out and a group of devices is targeted as a sex addict).

In the case of interpretable processes (translatable in human understandable language) a layer of human intervention is possible (although not necessary), as in, for example, the police search. Again this possibility is a double-edged sword since human intervention can either correct biases or insert new ones by interfering with the code, setting aside or inserting factors [95].

The lack of (legal and ethical) protocols to drive human action in designing and revising algorithms clearly calls for their creation, but requires a common setting of values and rules, since algorithms are basically enjoying a-territoriality in the sense that they are not necessarily used in one given physical jurisdiction. Moreover, the expansion of the autonomous generation of algorithms calls for building similar legal and ethical protocols to drive the machine generation of models. In both cases, there is also a need for technological verifiability of the effectiveness of the legal and ethical protocols making human readable at least the results of the application of the model when the model is not readable by humans.

Interpretable processes are certainly more accountable[62] and transparent[63] (although costlier),[64] but they still present risks. Moreover, by using (or translating into) interpretable processes, data mining will increase the possibility of getting back to searching for causality in the results instead of accepting a mere statistical association[65]: the group of devices passing through a city's solicitation block could be cleared of the social stigma the model has attached to them with all of its ensuing consequences. This is an important note since data mining results are NOT based

[61] For an explanation of the actual mechanisms see Solove [70].

[62] Meaning agents have an ethical and sometimes legal obligation to answer for their actions, wrongdoing, or mistakes.

[63] Transparency is intended as the enabling tool for actual accountability.

[64] The different cost-impact of the level of transparency required is analysed by Zarsky [52].

[65] Efforts to generate Transparency Enhancing Tools (TETs) is producing an expanding body of research at the crossroad between law and technology [96]. But on the side effects and risks of an excess of transparent information see Shkabatur [97].

on causality in the traditional sense but instead on mere probability.[66] Human readability would also enable a confidence check on the level of false positives in the rule produced by the algorithm, and, *ex post*, on the number of false negatives that were missed by the predictive model.[67]

What comes out of the preceding analysis is also a more nuanced notion of anonymity for the classifying society. After all, as clearly illustrated in the literature, anonymous profiles are "quantified identities imposed on individuals, often without their consultation, consent, or even awareness" [48, p. 1414].[68] The concept of anonymity[69] has influenced the very notion of personal data that is still defined in the EU GDPR as "*any information relating to an identified or identifiable natural person* ('data subject'); an identifiable natural person is one who can be identified, directly or indirectly, in particular by reference to an identifier such as a name, an identification number, location data, an online identifier, or one or more factors specific to the physical, physiological, genetic, mental, economic, cultural, or social identity of that natural person".[70] Meanwhile a "nothing-to-hide" cultural approach has been blooming with the expansion of the so-called web 2.0,[71] generating a less prudent approach to the data our devices generate and share. Most (not to say all) apps and software we use "require" access to (and also generate) an enormous amount of data that are unrelated to the purpose of the software or app. Those data are widely shared and subsequently fed in to a myriad of models that are then applied to the same and other sets of data-generating "things" (not necessarily to the individual owner of them).

Thus the technological promise of data protection through anonymization has been defeated by technology itself. Moreover, "... where anonymity for the sake of eliminating biases is desirable, one cannot assume that technical anonymity by itself guarantees the removal of bias. Rather, if technical anonymity allows for biased misattribution, then at least in some contexts, there may need to be additional steps introduced to mitigate this possibility" [105, pp. 178–179].

Finally, it is the very notion of pseudo-anonymous data[72] that provides the key to bypass entirely personal data protection in the classifying society.

---

[66]On this issue see in general Zarsky [22, 98].

[67]Literature concentrates on the potential harms of predictive algorithms [67].

[68]The "undiscovered observer represents the *quintessential* privacy harm because of the unfairness of his actions and the asymmetry between his and his victim's perspective" [89, p. 1160].

[69]The literature on the obstacles to obtaining acceptable levels of anonymity on the web is immense [39, 99–101]. See also Datalogix [102] privacy policy.

[70]Art. 4 GDPR.

[71]See the seminal work of Solove [103]; see also Zarsky [104].

[72]According to the EU GDPR (art. 4) "pseudonymisation' means the processing of personal data in such a manner that the personal data can no longer be attributed to a specific data subject without the use of additional information, provided that such additional information is kept separately and is subject to technical and organisational measures to ensure that the personal data are not attributed to an identified or identifiable natural person".

2.3 *Data and Markets: A Mismatch Between Law, Technology, and Businesses Practices*

A modern clash of rights seems to emerge in the classifying society—between, on the one hand, the right of individuals to control their own data, and, on the other, the interest of business in continuously harnessing that data[73] as an asset. The latter are increasingly protected by IP or quasi IP as trade secrets.[74]

This clash is echoed among economists. Some support business arguing that more (customers') data disclosure reduces information asymmetries [108, 109] while others argue it is data protection that generates social welfare [110]. In this debate legal scholars mostly advocate more individuals' control over their data at least in terms of an extended propertization [111, 112].[75] However, theories [113, 116, 117] about how a market for personal data, acknowledging data subjects' property rights, do not seem to consider the actual technological state of art [21] and the corresponding legal landscape.

In this landscape the legal framework remains dominated by the Privacy Policy Terms and Conditions whose legal enforceability remains at least doubtful in many legal orders. Meanwhile, *de facto* PPTC are enforced by the lack of market alternatives, by the widespread lack of data subjects' awareness both of the existence of tracking procedures and of the role of data aggregators and data brokers.[76] The hurdles and costs of a lawsuit close this vicious loop in which, although doubtful in legal terms, PPTC actually govern the data processing beyond statutory constraints in any given legal system.

In addition, the composite nature of most transactions makes opaque, to say the least, information collection and sharing processes. Very often, data processing is externalized [119] making it even more opaque and difficult to track data once they are shared.

Note also that some key players such as Google, Amazon, and e-Bay do not identify themselves as data brokers even though they regularly transfer data[77] and generate models.

Further and subsequent processing, even if data are anonymized or pseudo-anonymized, generates models that are later applied to the same (or other) groups of shared characteristics, a process which impacts individuals whose data, although technically non-personal, matches the model. In other words, the production cycle

[73]An average of 56 parties track activities on a website [106]. On the evolution of personal data trade see World Economic Forum [107].

[74]"No one can challenge the process of scoring and the results because the algorithms are zealously guarded trade secrets" [10, p. 5]. As illustrated by Richards and King [66, p. 42], "[w]hile Big Data pervasively collects all manner of private information, the operations of Big Data itself are almost entirely shrouded in legal and commercial secrecy".

[75]But see for concerns over propertization Noam [113]; Cohen [114]; Bergelson [115].

[76]The role of data aggregation and data brokers is vividly illustrated by Kroft [118].

[77]See e.g. Kosner [120].

we are describing (data collection of personal/things data, their enrichment with other data, and their pseudo-anonymization, the generation of the model and its continuous refinement) allows for application of the model to individuals without the need to actually involve further personal data (in legal technical terms), entirely bypassing data protection laws—even the most stringent ones.

Paradoxically, as it might seem that such a process could easily pay formal homage to a very strict reading of the necessity principle and the ensuing data minimization. Incidentally, the design of privacy policy terms that require (better yet: impose) consent in order to access the required services in a technological ecosystem that does not provide alternatives[78] trumps altogether those principles, while a business model based on the classifying society and on things' data (that is pseudo-anonymized data) disregards those principles altogether. In such a scenario, the data minimization rule normally required under the EU data protection directive[79] to distinguish between information needed to actually deliver the good or service and additional data becomes irrelevant—in any event, it has never acquired actual grip in business practice due to the low enforcement rate and despite its potential ability to increase data salience [119].

Technology experts have developed various concepts to enhance privacy protection—e.g. k-anonymity [123], l-diversity [124] and t-closeness [125]—but these concepts have revealed themselves as non-conclusive solutions. A similar fate is common to other privacy-enhancing technologies (PETs) [126, 127] that rely on the assumed preference for anonymity, clearly (yet problematically) challenged by the use of social networks,[80] while Do-Not-Track initiatives have not obtained enough success either [133]. And yet, the importance of all these attempts is becoming progressively less relevant in the classifying society. Accordingly, users must keep relying on regulators to protect them and on businesses to follow some form of ethical behaviour[81] calling for more appropriate answers from both stakeholders.

As mentioned, device fingerprints enable identification of individuals and devices themselves even when, for instance, cookies are disabled. If we consider that each device is composed of several other devices with their own digital fingerprints (e.g. a smart phone has a unique identification number while its Bluetooth and Wi-Fi antennas and other components also have their own), and that each of them continuously exchange information that can be compiled, the shift from personal

[78]Other authors have already pointed out that one key reading of privacy in the digital age is the lack of choice about the processes that involve us and the impossibility of understanding them [121, p. 133].

[79]See now the GDPR; for a technical analysis see Borcea-Pfitzmann et al. [122].

[80]See Gritzalis [128]. Indeed several authors have already highlighted the risks to privacy and autonomy posed by the expanding use of social networks: see, for instance the consequent call for a "Social Network Constitution" by Andrews [129] or the proposal principles of network governance by Mackinnon [82] or the worries expressed by Irani et al. [130, 131]; see also Sweeney [123]; Spiekermann et al. [132].

[81]See Fujitsu Res. Inst. [134].

data to things' data and to clusters of things' (non-personal) data as the subject of data processing begins to emerge in all its clarity and pervasiveness. When the GSM of a mobile phone allows geo-localization, we are used to thinking and behaving as if the system is locating the individual owner of the mobile phone. Yet technically it is not; it is locating the device and the data can be easily pseudo-anonymized. Once it becomes possible to target the device and the cluster(s) of data contingently related to it in a model, without the need to expressly target a specifically identified individual, the switch from personal data to things' data is complete, requiring an alternative approach that has not yet fully emerged.

The situation described above leads to a deadly stalemate of expanding asymmetry among players in markets and in society overall.

The overarching concept behind the classifying society is that individuals are no longer targetable as individuals but are instead selectively addressed for the way in which some clusters of data that they (technically, one or more of their devices) share with a given model fit in to the model itself. For instance, if many of our friends on a social network have bought a given book, chances are that we will do it as well; the larger the number of people in the network the more likely it is we will conform. What results is that individuals (or, rather, the given cluster of information) might be associated with this classification (e.g. high buying interest) and be treated accordingly—for instance, by being advertised a higher price (as happened for OS users versus Windows users).

Note that we normally do not even know such a classification model exists, let alone that it is used. Moreover, if engineered well from a legal standpoint, a data processing might not even amount to a *personal data* processing, preventing any meaningful application of data protection rules. On the other hand, the construction of models on a (very) large array of data that might or might not include ours is beyond our control and is normally protected as a business asset (e.g. as a trade secret).

In addition, since the matching between individuals and the model just requires a reduced amount of data, the need for further enriching the cluster of data with personal identifying information is progressively reduced. Thus, it becomes increasingly easy for the model to be uninterested in "identifying" us—instead contenting itself to target a small cluster of data related to a given device/thing (or ensemble of devices)—because all of the relevant statistical information is already incorporated in the model. The model need not identify us nor search for truths or clear causality as long as a given data cluster statistically shares a given amount of data with the model. Moreover, even if the model gets it wrong, it can learn (automatically with the help of AI and machine learning techniques) from the mistake, refining itself by adding/substituting/removing a subset of data or giving it a different weight that adjusts the statistical matching results: (big) data generate models and the use of models generates new data that feed the models and change them in an infinite creative loop whose human (and especially data subject) control is lost [7].

Such a deep change is unsettling for data protection because it undermines any legal and technological attempt (e.g. of anonymization techniques) to intervene.

Also, it can lead to unpleasant—even if unsought—results. The societal impact and risks for freedom of speech have been highlighted elsewhere. Yet, the selection of information and the tailoring of the virtual (and real) world in which we interact, even if not concerted among businesses,[82] reduce the chances of producing divergent behaviours with the result of actually reinforcing the model, on the one hand, and reducing the possibility of divergence even further on the other.[83]

Accordingly, technical solutions against privacy loss prove unhelpful if actions are based upon the correspondence of a small number of shared characteristics to a model. Indeed, privacy-preserving techniques, privacy-by-design, and other forms of technical protection for privacy all aim at reducing the amount of information available to prevent identification or re-identification of data [135][84] while personal identification as such is fading away in the encounter between technology and businesses models driven by algorithms.

This state of affairs calls for further investigation and a different research approach.

3 A Four-Layer Approach Revolving Around Privacy as a Framework

As anticipated, the data economy offers many advantages for enterprises, users, consumers, citizens, and public entities offering unparalleled scope for innovation, social connection, original problem-solving, and problem discovery using algorithms and machine learning (e.g. on emerging risks or their migration patterns such as in pandemic emergencies).[85] The actual unfolding of these possibilities requires the gathering and processing of data on both devices and individuals that raises divergent concerns [140] related to, on one hand, the will to share and participate (in social networks, for instance, or in public alerting systems in the case of natural disasters or pharmacovigilance), and, on the other hand, to the reliability of the organisations involved, regardless of their private or public nature.[86]

The complete pervasion of interactive, context-sensitive mobile devices along with the exponential growth in (supposedly) user-aware interactions with service providers in the domains of transportation and health-care or personal fitness,

[82]We are not discussing a science fiction conspiracy to control human beings but the actual side effects of the embrace of specific technological advancements with specific business models and their surrounding legal constraints.

[83]This holds true also when the code is verified or programmed by humans with the risk of embedding in it, even unintentionally, the biases of the programmer: "Because human beings program predictive algorithms, their biases and values are embedded into the software's instructions, known as the source code and predictive algorithms" [10, p. 4].

[84]See also Danezis [136].

[85]See also Bengtsson et al. [137]; Wesolowski et al. [138, 139].

[86]See also Wood and Neal [141]; Buttle and Burton [142]; Pew Research Centre [143]; Reinfelder [144].

for instance, dynamically generates new sources of information about individual behaviour or, in our framework, the behaviour of (their) devices—including personal preferences, location, historical data, and health-related behaviours.[87] It is already well known that in the context of mobile cells, Wi-Fi connections, GPS sensors, and Bluetooth, et cetera, several apps use location data to provide 'spatial aware' information. These apps allow users to check in at specific locations or venues, to track other users' movements and outdoor activities, and to share this kind of information [151]. Health-state measurement and monitoring capabilities are being incorporated into modern devices themselves, or provided through external devices in the form of smart watches, wearable clips, and bands.

The information provided can be employed for health care, for personal fitness, or in general for obtaining measurable information leading to marvellous potentials both in public and private use. Nevertheless, once cast in light of the blossoming role of algorithms in the classifying society this phenomenon contributes to the above-mentioned concerns, and calls for a clear, technologically-neutral regulatory framework that moves from privacy and relates to all legal fields involved [152], empowering balanced dealing between individual data subjects/users and organised data controllers.

The legal landscape of algorithm regulation requires that traditional approaches to privacy, conceived as one component of a multifaceted and unrelated puzzle of regulations, are transcended, regardless of the diversity of approaches actually taken in any given legal order (opt-in versus opt-out; market versus privacy as a fundamental right, ...). Only if data protection features at the centre of the regulatory process can it offer a comprehensive approach that does not overlook the link between social and economic needs, law and the technological implementation of legal rules, without underestimating the actual functioning of specific devices and algorithms-based business models.

The general trends of technology and business modeling described so far demonstrate the need to embed effective protection in existing legal data control tools (privacy as a fundamental right, for instance, or privacy by default/design) and the eventual need/opportunity to introduce *sui generis* forms of proprietary protection (personal data as commodities) to counterbalance the expanding protection of businesses in terms of both trade secrets and recognition of fundamental freedoms.[88]

In this framework, predictive algorithms are the first to raise concerns—especially in terms of consent, awareness (well beyond the notice model), and salience. Our privacy as readers and our ability to actually select what we want to read is an important issue that promptly comes to the fore when predictive algorithms are used [154].[89]

---

[87] See also Elkin-Koren and Weinstock [145]; FTC Staff Report [146–148]; Canadian Offices of the Privacy Commissioners [149]; Harris [150].

[88] See for instance Zhang v. Baidu.com, Inc (2014). But see, *e.g.*, Case C-131/12; Pasquale [153].

[89] On the privacy concerns and their social impact see Latar and Norsfors [155]; Ombelet and Morozov [156].

Nevertheless, using privacy as a framework reference for other areas of law in dealing with the classifying society would require an integrated strategy of legal innovation and technical solutions.

3.1 *Revise Interpretation of Existing Rules*

The first layer of this strategy would leverage the existing sets of rules.

In the actual context of algorithms-centrality in our societies at any relevant level, we claim that it is necessary to explore the potential interplay between the fundamental principles of privacy law and other legal rules such as unfair practices, competition, and consumer protection. For instance, although it may be based on individual consent, a practice that in reality makes it difficult for users to be aware of what data processing they are actually consenting to might be suspicious under any of the previously mentioned sets of legal rules. Business models and practices that segregate customers in a technically unnecessary way (e.g. by offering asymmetric internet connections to business and non-business clients), directing some categories to use the services in certain ways that make data collection easier and more comprehensive (e.g. through concentrated cloud-based services or asymmetric pricing) might appear suspicious under another set of regulations once privacy as a framework illustrates their factual and legal implications.

Furthermore, a business model that targets data clusters directly but refers to individuals only indirectly, using models generated by algorithms without clearly acknowledging this, can be questionable under several legal points of view.

Within this landscape dominated by opacity and warranting transparency and accountability, a 'privacy as a framework' approach is required to overcome a siloed legal description of the legal rules in order to cross-supplement the application of legal tools otherwise unrelated among each other. This layer of the approach envisions not unsettling regulatory changes by the rule makers but the use of traditional hermeneutic tools enlightened by the uptake of the algorithm-technological revolution.

In this vein some authors have begun to explore the application of the unconscionability doctrine to contracts that unreasonably favour data collecting companies [157].[90] To provide another example, once we tackle the topic of predictive algorithms, it is unavoidable to delve into the unfair commercial practices legal

[90]The authors also propose: "mandatory active choice between payment with money and payment with data, ex post evaluation of privacy notices, democratized data collection, and wealth or income-responsive fines". Their proposals enrich an already expanding host of regulatory suggestions. See Hajian and Domingo-Ferrer [158]; Mayer-Schonberger and Cukier [159]; Barocas and Selbst [19]. For a more technical account on fostering discrimination-free classifications, see Calders and Verwer [160]; Kamiran et al. [161]. Recently, the establishment of an ad hoc authority has also been advocated [162]. On market manipulation through the use of predictive and descriptive algorithms see the seminal work of Calo [38].

domain. All western societies devote a specific regulation to them.[91] In a March 2014 panel, the Federal Trade Commission [163] identified some topics relevant to our discussion[92]:

- "How are companies utilizing these predictive scores?
- How accurate are these scores and the underlying data used to create them?
- How can consumers benefit from the availability and use of these scores?
- What are the privacy concerns surrounding the use of predictive scoring?
- What consumer protections should be provided; for example, should consumers have access to these scores and the underlying data used to create them?"

Predictive algorithms decide on issues relevant to individuals "not because of what they've done, or what they will do in the future, but because inferences or correlations drawn by algorithms suggest they may behave in ways that make them poor credit or insurance risks, unsuitable candidates for employment or admission to schools or other institutions, or unlikely to carry out certain functions." The panel concluded by insisting on "transparency, meaningful oversight and procedures to remediate decisions that adversely affect individuals who have been wrongly categorized by correlation." The need to "ensure that by using big data algorithms they are not accidently classifying people based on categories that society has decided—by law or ethics—not to use, such as race, ethnic background, gender, and sexual orientation" was considered urgent [164, pp. 7–8].

Yet, if we are concerned about de-biasing the algorithms in order to avoid "already" non-permitted discriminations[93] and to promote transparency (intended both as accountability and as disclosure to customers),[94] we should be even more concerned about the risks posed by "arbitrariness-by-algorithm", which are more pressing for the million facets of our characteristics, either permanent or temporary—such as our momentary mood, or our movement through a specific neighbourhood—that the classifying society makes possible.

[91]See the EU Directive 2005/29/EC of the European Parliament and of the Council of 11 May 2000, concerning unfair business-to-consumer commercial practices in the internal market.

[92]We do not report here those topics that are exclusively related to the USA on which the FTC has authority, such as the relevance of the Fair Credit Reporting Act.

[93]For a recent account of algorithms' disparate impact see Barocas and Selbst [19, p. 671].

[94]See an analysis of some techniques potentially available to promote transparency and accountability [165]. See also Moss [24, p. 24], quoting relevant American statutes. Yet if action is not taken at a global level, online auditing can be run in countries where it is not forbidden and results transferred as information in other places. Analogously, a technical attempt to create auditing by using volunteers profiling in a sort of crowdsourcing empowering exercise might even make permissible online auditing in those mentioned jurisdictions forbidding the violation of PPT of web sites by using bots. There is an ongoing debate on this issue. See Benkler [166]; Citron [167]. But see *contra* Barnett [168]. For a critical analysis urging differentiation of the approach targeting the specific or general public see Zarsky [52].

Once we begin to ask ourselves whether it is fair is to take advantage of the vulnerabilities [169], of health-related[95] or other highly sensitive information [171], the need for a legal and ethical paradigm emerges vigorously and calls for reinterpreting the existing rules to expand their scope.

Indeed, once it is acknowledged that ToS and PPTCs are normally not read before accepting them [172],[96] and that the actual flow of data is often not even perceived by users because it runs in the background, the only apparent safeguard would seem to be an application of the necessity principle that not only limits data gathering and processing but that is normatively prescriptive well beyond authorizing a data processing.

For example, everybody realises that pre-installed apps[97] on ICT devices do not even allow activation of the apps themselves before consenting to the ToS and PPTCs, let alone the kind and extent of personal data collection and processing that they involve.[98] A sound application of the necessity principle (and of the privacy by design and by default approach to regulation) would impose a deep change in the design of web sites, of distributed products, of the structure of licence agreements, and of patterns of iterative bargaining even before dealing with the content of clauses on data processing and further uses of collected or given data.

For this reason, both the re-interpretation of existing legal rules and the eventual enactment of new ones are both necessary components of the overall strategy.

Such a process, which reverses the ongoing deregulation output via private law embedded in the content of the terms and in the design of their deployment in the progress of the business relationship (the pattern of alluring customers to buy), is rather intrusive on the part of the State and might be better pursued by reinterpreting existing private law rules governing the relationship between data subjects and data processors and/or service/product providers. This means enabling an interpretation of existing remedies (for instance against unfair business practices or of tort law rules, consumer protection statutes, and mistake and errors doctrines, ...) enlightened by the deep and subterranean change in the technological and legal landscape we have described so far.

For instance, if data protection rules are insufficient to prevent every installed app from claiming access to the full range of personal data present on a device in order to work, perhaps the (unknown or not read) terms requesting such an extension of access can be considered illicit—as unfair terms, unfair practices, or as anticompetitive, thereby triggering traditional contractual or liability remedies.

Reinterpretation, however, presents its own limits and certainly cannot cover every facet of the classifying society.

---

[95]On the potential for discriminatory and other misuses of health data regularly "protected" by professional secrecy see Orentlicher [170].

[96]Indeed, it has been estimated that on average we would need 244 h per year to read every privacy policy we encounter [173].

[97]Here, app is used as a synonym for software.

[98]See also Lipford et al. [174]; Passera and Haapio [175].

3.2 Promote Changes in Regulation and Empower Data Subjects Technologically

On a different level, the approach could lead to: (1) a reclassification of legal rules applicable when algorithms play a key role related to privacy along innovative lines; (2) the production of new rules in the same instances.

Such an approach would influence and inform the entire regulatory framework for both public and private activities, aiming at overcoming the pitfalls of an approach that maintains separate rules, for instance, on data protection, competition, unfair practices, civil liability, and consumer protection. However, to date, such an unsettling move is not realistically foreseeable.

Innovative regulation would be required, for example, in rearranging the protection of the algorithms as trade secrets in order to serve an emerging trend of considering personal data as (at least partially) property of the data subject, thereby enabling some forms of more explicit economic exchange. However, this innovation would also require the development and deployment of appropriate technology measures.

Data generation is not treated as an economic activity for the data subject while data gathering from data processors is characterized as such. This is mostly due to the fact that technologies to harness and exploit data from the business side are largely available while technology on the data subject side has to date failed to empower data subjects with workable tools for maintaining economic control over the information related to them.

The need to strengthen the legal and technological tools of data subjects due to the scale and pervasive use of algorithms in our societies is clearly manifest in the literature across any scientific field.

From our point of view, however, the background premise of data subject empowerment is similar to the one we can make on consumer protection, for instance. Both consumer protection and data protection laws endow consumers/data subjects with a large set of rights with the aim of reducing information asymmetries and equalizing bargaining power. The effective implementation of these consumer rights is mainly based on information that businesses are required to provide at an important cost for them. Indeed, both the relevant legal framework and the EUCJ consider consumers to be rational individuals capable of becoming "reasonably well-informed and reasonably observant and circumspect" (Case C-210/96, par. 31), for instance by routinely reading food labels and privacy policies (Case C–51/94),[99] and of making efficient decisions so long as the relevant (although copious) information is presented to them in "a comprehensible manner".[100]

[99] See Case C–51/94, para 34, holding that consumers who care about ingredients (contained in a sauce) read labels (*sic*); see also Phillips [176]; Gardner [177]; Ayres and Schwartz [178].

[100] E.g. artt. 5, 6 and 10 of Directive 2000/31/EC of the European Parliament and of the Council of 8 June 2000 on certain legal aspects of information society services, in particular electronic commerce, in the Internal Market ('Directive on electronic commerce').

However, proposing to provide information in order to remedy bargaining asymmetries and sustain a proper decision-making process does not seem to effectively fill the gaps between formal rules and their impact on the daily life of consumers/users.[101] Once information is technically provided[102] and contract/TOS/PPTCs[103] are formally agreed upon, customers remain trapped in them as if they had actually profited from mandated information transparency by having freely and rationally decided upon such terms. Despite all the costs for industries, the attempt to reduce asymmetries and to empower efficient consumer rights often remains a form of wishful thinking[104] with a significant impact on consumer trust and market growth.

Indeed, very few people read contracts when buying products or services online [172]. Actually, the expectation that they would do so is possibly unrealistic, as a typical individual would need, on average, 244 h per year just to read every privacy policy they encounter [173], let alone the overall set of information that is actually part of the contract and the contract itself [172].

Moreover, it is likely that users are not encouraged to read either terms and conditions or information related to the product/service due to systematic patterns in language and typographic style that contribute to the perception of such terms as mere document formalities [187] and/or make them uneasily accessible (see, for example, actual difficulties in reading labels and information displayed on packages).

The difficulty of reading terms and technical information stems out not only from their surface structure but also from readers' lack of familiarity with the underlying concepts [188] and the fact that they figure in a format totally unfamiliar and not plainly understandable to the majority of consumers. Moreover, the vagueness and complex language of clauses, along with their length, stimulate inappropriate and inconsistent decision patterns on behalf of users.

As a result, lack of effective information, unawareness, and perception of a loose control over the process of setting preferences have a decisive impact on consumer trust, leading to weakening consumption patterns and to expanding asymmetries of power between individual users/customers/consumers and businesses.

---

[101] See also Bar-Gill and Ben-Shahar [179]; Luzak [180, 181]; Purnhagen and Van Herpen [182].

[102] See for an information mandate approach: Council Directive 93/13/EEC of 5 April 1993 on unfair terms in consumer contracts; Directive on electronic commerce; Directive 2005/29/EC of the European Parliament and of the Council of 11 May 2005 concerning unfair business-to-consumer commercial practices in the internal market; Directive 2008/48/EC of the European Parliament and of the Council of 23 April 2008 on credit agreements for consumers and repealing Council Directive 87/102/EEC; Directive 2011/83/EU of the European Parliament and of the Council of 25 October 2011 on consumer rights and Regulation (Eu) No 531/2012 of the European Parliament and of the Council of 13 June 2012 on roaming on public mobile communications networks within the Union.

[103] See McDonald et al. [183].

[104] See in the economics literature: Stigler [184]; Akerlof [185]; Macho-Stadler and Perez-Castrillo [186].

Reversing this state of affairs to actually empower users/data subjects to select and manage the information they want (and want to share), and to expand their bargaining powers as a consequence of their ability to govern the flow of information, requires a deep change in approach. A change as deep as it is simple. Since businesses are not allowed to change the notice and consent approach[105] (although they largely manipulate it), a careful selection of useful information can be construed only from the other end (data subject/users).

The concept is rather simple: law forces the revelation of a large amount of information useful to users/data subjects, but the latter need an application tool to efficiently manage such information in order to actually become empowered and free in their choices.[106] While businesses cannot lawfully manage to reduce the amount of information they owe to users as consumers or data subjects, users can reduce and select that information for their improved use and for a more meaningful comparison in a functioning market.

A key feature of this layer in our approach is to embrace, technologically and in regulatory terms, a methodology pursuing an increased technological control over the flow of information, also with the assistance of legal rules that sustain such a control.

Yet, we are aware of the difficulties this layer entails. Indeed, it is difficult to regain a degree of control over a large part of data used by algorithms or by the models they create. Indeed, as off-line customers, our data are collected and processed either by relating them to devices we wear (e.g. e-objects), carry (e.g. smart phones), or use (e.g. fidelity cards, credit cards, apps, ...) when buying products or services (e.g. paying the highway or the parking with electronic means).

It is important to note that a personal data-safe technological approach (e.g. data are maintained in the control of the data subjects that licence access to them) would require that businesses actually consent to opt-in to a legal system that allows and enforces the potential for data subjects to avoid surrendering data beyond a clear and strict application of the necessity principle. Similarly, and where algorithms' functioning remains unveiled by the shadow of trade secrets protection, only a technological and cooperative approach can be of help. For instance, collective sharing of experiences along with shared (high) computational abilities can "reverse engineer" algorithms' outputs to detect old and new biased results.

In other words, without a strict enforcement (that requires both a technological and a legal grip) of the necessity principle in delivering goods and services, we will never see a different market for data emerge—and certainly not one in which data subjects get most of the economic benefits from data (even pseudo-anonymous) related to them.

---

[105]On the failure of the disclosure model see in general Ben-Shahar and Schneider [189], Radin [190], and with reference to privacy Ben-Shahar and Chilton [191].

[106]The issue of actual market alternatives is not addressed here.

3.3 *Embed Ethical Principles in Designing Algorithms and Their Related Services*

There are several reasons to attempt to embed ethical principles in the design of algorithms and their related services [192, 193].

First of all, to date any technological and legal strategy to rebalance the allocation of information, knowledge, and power between users/data subjects and businesses/data processors and brokers has proved unsuccessful.[107]

The reinterpretation of the solutions in any given legal system in light of technological developments and a sound privacy-as-a-framework approach would require a strong commitment by scholars, courts, and administrative authorities. In any event, it would require a certain amount of time to deploy its effects.

To the contrary, ethical principles and social responsibility are already emerging as drivers of businesses' decision-making. Ethical principles, once technologically incorporable, in designing and testing algorithms and their related services can use exactly the same technology used by algorithms to monitor algorithms' and business models' actual operation.

Also, the adoption of such principles by the community of code developers, for instance, does not in principle need to receive the approval of businesses since these would be ethical rules of the "profession".

4 A Summary as a Conclusion

In the classifying society a new target for protection emerges: personal anonymities, the clusters of characteristics shared with an algorithm-generated model that are not necessarily related to personal identifiable information and thus are not personal data in the strict meaning even of EU law.

The centrality of personal anonymities is blossoming with the maturing of the Internet of Things that exponentially increases the potential of algorithms and of their use unrelated to personal identifying information.

Personal anonymities stem from innovative business models and applied new technologies. In turn, they require a combined regulatory approach that blends together (1) the reinterpretation of existing legal rules in light of the central role of privacy in the classifying society; (2) the promotion of disruptive technologies for disruptive new business models enabling more market control by data subjects over their own data; and, eventually, (3) new rules aiming, among other things, to provide to data generated by individuals some form of property protection similar to that enjoyed by the generation of data and models by businesses (e.g. trade secrets).

---

[107] See above footnotes 80–81 and accompanying text.

The blend would be completed by (4) the timely insertion of ethical principles in the very generation of the algorithms sustaining the classifying society.

Different stakeholders are called to intervene in each of the above mentioned layers of innovation.

None of these layers seems to have a prevalent leading role, to date. However, perhaps the technical solutions enabling a market for personal data led by data subjects, if established, would catalyse the possibility of generating alternate business models more in line with the values reflected in personal data protection rules. Yet, the first step remains to put privacy at the centre of the wide web of legal rules and principles in the classifying society.

References

1. European Data Protection Supervisor, Opinion No 4/2015: Towards a new digital ethics: Data, dignity and technology. https://secure.edps.europa.eu/EDPSWEB/webdav/site/mySite/shared/Documents/Consultation/Opinions/2015/15-09-11_Data_Ethics_EN.pdf. Accessed 24 Oct 2016
2. Angwin, J.: The web's new gold mine: your secrets. Wall Street J. http://online.wsj.com/article/SB10001424052748703940904575395073512989404.html (2010). Accessed 24 Oct 2016
3. Bain & Company: Using Data as a Hidden Asset. http://www.bain.com/publications/articles/using-data-as-a-hidden-asset.aspx (2010). Accessed 24 Oct 2016
4. Pariser, E.: The Filter Bubble. Penguin Press, New York (2011)
5. Tene, O., Polonetsky, J.: Big data for all: privacy and user control in the age of analytics. Northwest. J. Technol. Intellect. Prop. **11**(5), 239–273 (2013)
6. Ohm, P.: Response, the underwhelming benefits of big data. Univ. Pa. Law Rev. Online. **161**, 339–346 (2013)
7. Van Otterlo, M.: Automated experimentation in Walden 3.0: the next step in profiling, predicting, control and surveillance. Surveill. Soc. **12**(2), 255–272 (2014)
8. Lomas, N.: Amazon patents "anticipatory" shipping—to start sending stuff before you've bought it. https://techcrunch.com/2014/01/18/amazon-pre-ships/ (2014). Accessed 24 Oct 2016
9. Vaidhyanathan, S.: The Googlization of Everything. University of California Press, Berkeley (2011)
10. Citron, D.K., Pasquale, F.: The scored society: due process for automated predictions. Wash. Law Rev. **89**(1), 1–33 (2014)
11. Duhigg, C.: How Companies Learn Your Secrets. The New York Times, New York (2012). http://www.nytimes.com/2012/02/19/magazine/shopping-habits.html. Accessed 24 Oct 2016.
12. Gray, D., Citron, D.K.: The right to quantitative privacy. Minn. Law Rev. **98**, 62–144 (2013)
13. Schwartz, P.M.: Privacy and democracy in cyberspace. Vanderbilt Law Rev. **52**, 1609–1701 (1999)
14. Schwartz, P.M.: Internet privacy and the state. Conn. Law Rev. **32**, 815–859 (2000)
15. Cohen, J.E.: Examined lives: informational privacy and the subject as object. Stanford Law Rev. **52**, 1373–1438 (2000)
16. Cohen, J.E.: Cyberspace as/and space. Columbia Law Rev. **107**(1), 210–256 (2007)
17. Solove, D.J.: The Digital Person. New York University Press, New York (2004)

18. Mattioli, D.: On Orbitz, Mac users steered to pricier hotels. Wall Street J.. **23**, 2012 (2012). http://www.wsj.com/articles/SB10001424052702304458604577488822667325882. Accessed 24 Oct 2016.
19. Barocas, S., Selbst, A.D.: Big data's disparate impact. Calif. Law Rev. **104**, 671–732 (2016)
20. Colb, S.F.: Innocence, privacy, and targeting in fourth amendment jurisprudence. Columbia Law Rev. **56**, 1456–1525 (1996)
21. Korff, D.: Data protection laws in the EU: the difficulties in meeting the challenges posed by global social and technical developments. In: European Commission Directorate-General Justice, Freedom and Security, Working Paper No. 2. http://ec.europa.eu/justice/policies/privacy/docs/studies/new_privacy_challenges/final_report_working_paper_2_en.pdf (2010). Accessed 24 Oct 2016
22. Zarsky, T.Z.: Governmental data mining and its alternatives. Penn State Law Rev. **116**(2), 285–330 (2011)
23. Bamberger, K.A.: Technologies of compliance: risk and regulation in a digital age. Tex. Law Rev. **88**(4), 669–739 (2010)
24. Moss, R.D.: Civil rights enforcement in the era of big data: algorithmic discrimination and the computer fraud and abuse act. Columbia Hum. Rights Law Rev. **48**(1) (2016).
25. Exec. Office of The President: Big data: seizing opportunities, preserving values. http://www.whitehouse.gov/sites/default/files/docs/big_data_privacy_report_5.1.14_final_print.pdf (2014). Accessed 24 Oct 2016
26. Turow, J.: Niche Envy. MIT Press, Cambridge, MA (2006)
27. Al-Khouri, A.M.: Data ownership: who owns "my data"? Int. J. Manag. Inf. Technol. **2**(1), 1–8 (2012)
28. Rajagopal, S.: Customer data clustering using data mining technique. Int. J. Database Manag. Syst. **3**(4), 1–11 (2011)
29. Frischmann, B.M., Selinger, E.: Engineering humans with contracts. Cardozo Legal Studies Research Paper No. 493. https://ssrn.com/abstract=2834011 (2016). Accessed 24 Oct 2016
30. Perzanowski, A., Hoofnagle, C.J.: What we buy when we 'buy now'. Univ. Pa. Law Rev. **165**, 317 (2017). https://papers.ssrn.com/sol3/papers.cfm?abstract_id=2778072 (forthcoming 2017). Accessed 24 Oct 2014
31. Apple: Terms and Conditions—Game Center. http://www.apple.com/legal/internet-services/itunes/gamecenter/us/terms.html (2013). Accessed 24 Oct 2016
32. Behm, R.: What are the issues? Employment testing: failing to make the grade. http://employmentassessment.blogspot.com/2013/07/what-are-issues.html (2013). Accessed 24 Oct 2016
33. EFF: Timeline of NSA domestic spying. https://www.eff.org/nsa-spying/timeline (2015). Accessed 24 Oct 2016
34. Schneier, B.: Want to evade NSA spying? Don't connect to the internet. Wired Magazine. http://www.wired.com/opinion/2013/10/149481 (2013). Accessed 24 Oct 2016
35. Rosenbush, S.: Facebook tests software to track your cursor on screen. CIO J. http://blogs.wsj.com/cio/2013/10/30/facebook-considers-vast-increase-in-data-collection (2013). Accessed 24 Oct 2016
36. PrivacySOS: NYPD's domain awareness system raises privacy, ethics issues. https://privacysos.org/blog/nypds-domain-awareness-system-raises-privacy-ethics-issues/ (2012). Accessed 24 Oct 2016
37. TrapWire: The intelligent security method. http://www.trapwire.com/trapwire.html (2016). Accessed 24 Oct 2016
38. Calo, M.R.: Digital market manipulation. George Wash. Law Rev. **82**(4), 95–1051 (2014)
39. Ohm, P.: Broken promises of privacy: responding to the surprising failure of anonymization. UCLA Law Rev. **57**, 1701–1777 (2010)
40. Hoofnagle, C.Y.: Big brother's little helpers: how choicepoint and other commercial data brokers collect and package your data for law enforcement. N. C. J. Int. Law Commer. Regul. **29**, 595–637 (2004)

41. Singer, N.: Mapping, and sharing, the consumer genome. The New York Times. http://www.nytimes.com/2012/06/17/technology/acxiom-the-quiet-giant-of-consumer-database-marketing.html (2012). Accessed 24 Oct 2016
42. Tucker, P.: Has big data made anonymity impossible? MIT Technology Review. http://www.technologyreview.com/news/514351/has-big-data-madeanonymity-impossible/ (2013). Accessed 24 Oct 2016
43. Article 29 Data Protection Working Party: Opinion 5/2014 on anonymization techniques. http://ec.europa.eu/justice/data-protection/article-29/documentation/opinion-recommendation/files/2014/wp216_en.pdf (2014). Accessed 24 Oct 2016
44. Ruggieri, S., Pedreschi, D., Turini, F.: Data mining for discrimination discovery. ACM Trans. Knowl. Discov. Data. **4**(2), 1–40 (2010)
45. Hoofnagle, C.Y., Whittington, J.: "Free": accounting for the costs of the Internet's most popular price. UCLA Law Rev. **61**, 606–670 (2014)
46. Aziz, A., Telang, R.: What is a digital cookie worth? https://papers.ssrn.com/sol3/papers.cfm?abstract_id=2757325 (2016). Accessed 24 Oct 2016
47. Bozdag, E.: Bias in algorithmic filtering and personalization. Ethics Inf. Technol. **15**(3), 209–227 (2013)
48. Pasquale, F., Citron, D.K.: Promoting innovation while preventing discrimination: policy goals for the scored society. Wash. Law Rev. **89**(4), 1413–1424 (2014)
49. Pasquale, F.: Beyond innovation and competition: the need for qualified transparency in Internet intermediaries. Northwest. Univ. Law Rev. **104**(1), 105–174 (2010)
50. Pasquale, F.: Restoring transparency to automated authority. J. Telecommun. High Technol. Law. **9**, 235–256 (2011)
51. Zarsky, T.Z.: Thinking outside the box: considering transparency, anonymity, and pseudonymity as overall solutions to the problems in information privacy in the Internet society. Univ. Miami Law Rev. **58**, 1301–1354 (2004)
52. Zarsky, T.Z.: Transparent predictions. Univ. Ill. Law Rev. **2013**(4), 1503–1570 (2013)
53. Cohen, J.E.: Configuring the Networked Self: Law, Code, and the Play of Everyday Practice. Yale University Press, New Haven, CT (2012)
54. Citron, D.K., Gray, D.: Addressing the harm of total surveillance: a reply to professor Neil Richards. Harv. L. Rev. F. **126**, 262 (2013)
55. DHS: National network of fusion centers fact sheet. https://www.dhs.gov/national-network-fusion-centers-fact-sheet (2008). Accessed 24 Oct 2016
56. Cohen, J.E.: Privacy, visibility, transparency, and exposure. Univ. Chicago Law Rev. **75**(1), 181–201 (2008)
57. Karabenick, S.A., Knapp, J.R.: Effects of computer privacy on help-seeking. J. Appl. Soc. Psychol. **18**(6), 461–472 (1988)
58. Peck, D.: They're watching you at work. The Atlantic. http://www.osaunion.org/articles/Theyre%20Watching%20You%20At%20Work.pdf (2013). Accessed 24 Oct 2016
59. Citron, D.K.: Data mining for juvenile offenders. Concurring Opinions. http://www.concurringopinions.com/archives/2010/04/data-mining-for-juvenile-offenders.html (2010). Accessed 24 Oct 2016
60. Coleman, E.G.: Coding Freedom. Princeton University Press, Princeton (2013)
61. Marwick, A.E.: How your data are being deeply mined. The New York Review of Books. http://www.nybooks.com/articles/2014/01/09/how-your-data-are-being-deeply-mined/ (2014). Accessed 24 Oct 2016
62. Abdou, H.A., Pointon, J.: Credit scoring, statistical techniques and evaluation criteria: a review of the literature. Intell. Syst. Account. Finance Manag. **18**(2–3), 59–88 (2011)
63. Balkin, J.M.: The constitution in the national surveillance state. Minn. Law Rev. **93**(1), 1–25 (2008)
64. Kerr, O.S.: Searches and seizures in a digital world. Harv. Law Rev. **119**(2), 531–585 (2005)
65. Citron, D.K.: Technological due process. Wash. Univ. Law Rev. **85**(6), 1249–1313 (2008)
66. Richards, N.M., King, J.H.: Three paradoxes of big data. Stanford Law Rev. **66**, 41–46 (2013)

67. Crawford, K., Schultz, J.: Big data and due process: toward a framework to redress predictive privacy harms. Boston Coll. Law Rev. **55**(1), 93–128 (2014)
68. Ramasastry, A.: Lost in translation? Data mining, national security and the "adverse inference" problem. Santa Clara Comput. High Technol. Law J. **22**(4), 757–796 (2004)
69. Slobogin, C.: Government data mining and the fourth amendment. Univ. Chicago Law Rev. **75**(1), 317–341 (2008)
70. Solove, D.J.: Data mining and the security-liberty debate. Univ. Chicago Law Rev. **75**, 343–362 (2008)
71. Solove, D.J.: Privacy and power: computer databases and metaphors for information privacy. Stanford Law Rev. **53**(6), 1393–1462 (2001)
72. Cate, F.H.: Data mining: the need for a legal framework. Harv. Civil Rights Civil Liberties Law Rev. **43**, 435 (2008)
73. Strandburg, K.J.: Freedom of association in a networked world: first amendment regulation of relational surveillance. Boston Coll. Law Rev. **49**(3), 741–821 (2008)
74. Bloustein, E.J.: Individual and Group Privacy. Transaction Books, New Brunswick, NJ (1978)
75. Conseil National Numerique, Platform Neutrality: Building an open and sustainable digital environment. http://www.cnnumerique.fr/wp-content/uploads/2014/06/ PlatformNeutrality_VA.pdf (2014). Accessed 24 Oct 2016
76. Nunez, M.: Senate GOP launches inquiry into Facebook's news curation. http://gizmodo.com/senate-gop-launches-inquiry-into-facebook-s-news-curati-1775767018 (2016). Accessed 24 Oct 2016
77. Chan, C.: When one app rules them all: the case of WeChat and mobile in China. Andreessen Horowitz. http://a16z.com/2015/08/06/wechat-china-mobile-first/ (2015). Accessed 24 Oct 2016
78. ADL: Google search ranking of hate sites not intentional. http://archive.adl.org/rumors/google_search_rumors.html (2004). Accessed 24 Oct 2016
79. Woan, T.: Searching for an answer: can Google legally manipulate search engine results? Univ. Pa. J. Bus. Law. **16**(1), 294–331 (2013)
80. Wu, T.: Machine speech. Univ. Pa. Law Rev. **161**, 1495–1533 (2013)
81. Volokh, E., Falk, D.: First amendment protection for search engine search results. http://volokh.com/wp-content/uploads/2012/05/SearchEngineFirstAmendment.pdf (2012). Accessed 24 Oct 2016
82. MacKinnon, R.: Consent of the Networked. Basic Books, New York (2012)
83. Chander, A.: Facebookistan. N. C. Law Rev. **90**, 1807 (2012)
84. Pasquale, F.: Search, speech, and secrecy: corporate strategies for inverting net neutrality debates. Yale Law and Policy Review. Inter Alia. http://ylpr.yale.edu/inter_alia/search-speech-and-secrecy-corporate-strategies-inverting-net-neutrality-debates (2010). Accessed 24 Oct 2016
85. Richtel, M.: I was discovered by an algorithm. The New York Times. http://archive. indianexpress.com/news/i-was-discovered-by-an-algorithm/1111552/ (2013). Accessed 24 Oct 2016
86. Slobogin, C.: Privacy at Risk. University of Chicago Press, Chicago (2007)
87. Zarsky, T.Z.: Understanding discrimination in the scored society. Wash. Law Rev. **89**, 1375–1412 (2014)
88. Nissenbaum, H.F.: Privacy in Context. Stanford Law Books, Stanford, CA (2010)
89. Calo, M.R.: The boundaries of privacy harm. Indiana Law J. **86**(3), 1131–1162 (2011)
90. Goldman, E.: Data mining and attention consumption. In: Strandburg, K., Raicu, D. (eds.) Privacy and Technologies of Identity. Springer Science + Business Media, New York (2005)
91. Pasquale, F.: The Black Box Society: The Secret Algorithms That Control Money and Information. Harvard University Press, Cambridge, MA (2015)
92. Clarke, R.: Profiling: a hidden challenge to the regulation of data surveillance. J. Law Inf. Sci. **4**(2), 403 (1993)

93. Fayyad, U.M., Piatetsky-Shapiro, G., Smyth, P.: From data mining to knowledge discovery: an overview. In: Fayyad, U. (ed.) Advances in Knowledge Discovery and Data Mining. AAAI Press, Menlo Park, CA (1996)
94. Paparrizos, J., White, R.W., Horvitz, E.: Screening for pancreatic adenocarcinoma using signals from web search logs: feasibility study and results. J. Oncol. Pract. **12**(8), 737–744 (2016)
95. Friedman, B., Nissenbaum, H.: Bias in computer systems. ACM Trans. Inf. Syst. **14**(3), 330–347 (1996). In: Friedman, B. (ed.). Human Values and the Design of Computer Technology. CSLI Publications, Stanford, CA (1997)
96. Hildebrant, M.: Profiling and the rule of law. Identity Inf. Soc. **1**(1), 55–70 (2008)
97. Shkabatur, J.: Cities @ crossroads: digital technology and local democracy in America. Brooklin Law Rev. **76**(4), 1413–1485 (2011)
98. Zarsky, T.Z.: "Mine your own business!": making the case for the implications of the data mining of personal information in the forum of public opinion. Yale J. Law Technol. **5**(1), 1–56 (2003)
99. Mayer, J: Tracking the trackers: where everybody knows your username. http://cyberlaw.stanford.edu/node/6740 (2011). Accessed 24 Oct 2016
100. Narayanan, A: There is no such thing as anonymous online tracking. http://cyberlaw.stanford.edu/node/6701 (2011). Accessed 24 Oct 2016
101. Perito, D., Castelluccia, C., Kaafar, M.A., Manilsr, P.: How unique and traceable are usernames? In: Fischer-Hübner, S., Hopper, N. (eds.) Privacy Enhancing Technologies. Springer, Berlin (2011)
102. Datalogix: Privacy policy. https://www.datalogix.com/privacy/ (2016). Accessed 24 Oct 2016
103. Solove, D.J.: Nothing to Hide. Yale University Press, New Haven, CT (2011)
104. Zarsky, T.Z.: Law and online social networks: mapping the challenges and promises of user-generated information flows. Fordham Intell. Prop. Media Entertainment Law J. **18**(3), 741–783 (2008)
105. Himma, K.E., Tavani, H.T.: The Handbook of Information and Computer Ethics. Wiley, Hoboken, NJ (2008)
106. Angwin, J.: Online tracking ramps up—popularity of user-tailored advertising fuels data gathering on browsing habits. Wall Street J. http://www.wsj.com/articles/SB10001424052702303836404577472491637833420 (2012). Accessed 24 Oct 2016
107. World Economic Forum: Rethinking personal data: strengthening trust. http://www3.weforum.org/docs/WEF_IT_RethinkingPersonalData_Report_2012.pdf (2012). Accessed 24 Oct 2016
108. Posner, R.A.: The economics of privacy. Am. Econ. Rev. **71**(2), 405–409 (1981)
109. Calzolari, G., Pavan, A.: On the optimality of privacy in sequential contracting. J. Econ. Theory. **130**(1), 168–204 (2006)
110. Acquisti, A., Varian, H.R.: Conditioning prices on purchase history. Mark. Sci. **24**(3), 367–381 (2005)
111. Schwartz, P.M.: Property, privacy, and personal data. Harv. Law Rev. **117**, 2056–2128 (2003)
112. Purtova, N.: Property rights in personal data: an European perspective. Dissertation, Uitgeverij BOXPress, Oistervijk (2011)
113. Noam, E.M.: Privacy and self-regulation: markets for electronic privacy. In: Wellbery, B.S. (ed.) Privacy and Self-Regulation in the Information Age. U.S. Dept. of Commerce, National Telecommunications and Information Administration, Washington, D.C. (1997)
114. Cohen, J.E.: Examined lives: informational privacy and the subject as object. Stanford Law Rev. **52**, 1373–1437 (1999)
115. Bergelson, V.: It's personal but is it mine? Toward property rights in personal information. U.C. Davis Law Review. **37**, 379–451 (2003)
116. Laudon, K.C.: Markets and privacy. Commun. ACM. **39**(9), 92–104 (1996)
117. Aperjis, C., Huberman, B.: A market for unbiased private data: paying individuals according to their privacy attitudes. First Monday **17**(5) (2012)

118. Kroft, S.: The data brokers: selling your personal information. 60 Minutes. http://www.cbsnews.com/news/data-brokers-selling-personal-information-60-minutes/ (2014). Accessed 24 Oct 2016
119. Jentzsch, N., Preibusch, S., Harasser, A.: Study on monetizing privacy: an economic model for pricing personal information. ENISA Publications. https://www.enisa.europa.eu/publications/monetising-privacy (2012). Accessed 24 Oct 2016
120. Kosner, A.W.: New Facebook policies sell your face and whatever it infers. Forbes. http://www.forbes.com/sites/anthonykosner/2013/08/31/new-facebook-policies-sell-your-faceand-whatever-it-infers/ (2013). Accessed 24 Oct 2016
121. Solove, D.J.: Understanding Privacy. Harvard University Press, Cambridge, MA (2008)
122. Borcea-Pfitzmann, K., Pfitzmann, A., Berg, M.: Privacy 3.0: = data minimization + user control + contextual integrity. Inf. Technol. **53**(1), 34–40 (2011)
123. Sweeney, L.: K-anonymity: a model for protecting privacy. Int. J. Uncertain Fuzziness Knowl Based Syst. **10**(5), 557–570 (2002)
124. Machanavajjhala, A., Kifer, D., Gehrke, J., Venkitasubramaniam, M.: L-diversity: privacy beyond k-anonymity. ACM Trans. Knowl. Discov. Data **1**(1), 1–52, Art. 3 (2007)
125. Li, N., Li, T., Venkatasubramanian, S.: t-closeness: privacy beyond k-anonymity and l-diversity. In: IEEE 23rd International Conference on Data Engineering, pp. 106–115. IEEE, Istanbul (2007)
126. Karjoth, G., Schunter, M., Waidner, M.: Privacy-enabled services for enterprises. http://www.semper.org/sirene/publ/KaSW_02.IBMreport-rz3391.pdf (2002). Accessed 24 Oct 2016
127. Cranor, L.F., Guduru, P., Arjula, M.: User interfaces for privacy agents. ACM Trans. Comput. Hum. Interact. **13**(2), 135–178 (2006)
128. Gritzalis, S.: Enhancing web privacy and anonymity in the digital era. Inf. Manag. Comput. Secur. **12**(3), 255–288 (2004)
129. Andrews, L.: I Know Who You Are and I Saw What You Did: Social Networks and The Death of Privacy. Free Press, New York (2012)
130. Irani, D., Webb, S., Li, K., Pu, C.: Large online social footprints—an emerging threat. http://cobweb.cs.uga.edu/~kangli/src/SecureCom09.pdf (2009). Accessed 24 Oct 2016
131. Irani, D., Webb, S., Pu, C., Li, K.: Modeling unintended personal-information leakage from multiple online social networks. IEEE Internet Comput. **15**(3), 13–19 (2011)
132. Spiekermann, S., Dickinson, I., Günther, O., Reynolds, D.: User agents in e-commerce environments: industry vs. consumer perspectives on data exchange. In: Eder, J., Missikoff, M. (eds.) Advanced Information Systems Engineering. Springer, Berlin (2003)
133. Bott, E.: The do not track standard has crossed into crazy territory. http://www.zdnet.com/the-do-not-track-standard-has-crossed-into-crazy-territory-7000005502/ (2012). Accessed 24 Oct 2016
134. Fujitsu Res. Inst.: Personal data in the cloud: a global survey of consumer attitudes. http://www.fujitsu.com/downloads/SOL/fai/reports/fujitsu_personal-data-in-the-cloud.pdf (2010). Accessed 24 Oct 2016
135. Brunton, F., Nissenbaum, H.: Vernacular resistance to data collection and analysis: a political theory of obfuscation. First Monday. **16**(5), 1–16 (2011)
136. Danezis, G.: Privacy technology options for smart metering. http://research.microsoft.com/enus/projects/privacy_in_metering/privacytechnologyoptionsforsmartmetering.pdf (2011). Accessed 24 Oct 2016
137. Bengtsson, L., Lu, X., Thorson, A., Garfield, R., von Schreeb, J.: Improved response to disasters and outbreaks by tracking population movements with mobile phone network data: a post-earthquake geospatial study in Haiti. PLoS Med. **8**(8), e1001083 (2011)
138. Wesolowski, A., Eagle, N., Tatem, A.J., Smith, D.L., Noor, A.M., Snow, R.W., Buckee, C.O.: Quantifying the impact of human mobility on malaria. Science. **338**(6104), 267–270 (2012)

139. Wesolowski, A., Buckee, C., Bengtsson, L., Wetter, E., Lu, X., Tatem, A.J.: Commentary: containing the ebola outbreak—the potential and challenge of mobile network data. http://currents.plos.org/outbreaks/article/containing-the-ebola-outbreak-the-potential-and-challenge-of-mobile-network-data/ (2014). Accessed 24 Oct 2016
140. Phelps, J., Nowak, G., Ferrell, E.: Privacy concerns and consumer willingness to provide personal information. J. Public Policy Mark. **19**(1), 27–41 (2000)
141. Wood, W., Neal, D.T.: The habitual consumer. J. Consum. Psychol. **19**(4), 579–592 (2009)
142. Buttle, F., Burton, J.: Does service failure influence customer loyalty? J. Consum. Behav. **1**(3), 217–227 (2012)
143. Pew Research Centre: Mobile health 2012. http://www.pewinternet.org/2012/11/08/mobile-health-2012 (2012). Accessed 24 Oct 2016
144. Reinfelder, L., Benenson, Z., Gassmann, F.: Android and iOS users' differences concerning security and privacy. In: Mackay, W. (ed.) CHI '13 Extended Abstracts on Human Factors in Computing Systems. ACM, New York, NY (2013)
145. Elkin-Koren, N., Weinstock Netanel, N. (eds.): The Commodification of Information. Kluwer Law International, The Hague (2002)
146. FTC Staff Report: Mobile apps for kids: current privacy disclosures are disappointing. http://www.ftc.gov/os/2012/02/120216mobile_apps_kids.pdf (2012). Accessed 24 Oct 2016
147. FTC Staff Report: Mobile apps for kids: disclosures still not making the grade. http://www.ftc.gov/os/2012/12/121210mobilekidsappreport.pdf (2012). Accessed 24 Oct 2016
148. FTC Staff Report: Mobile privacy disclosures: building trust through transparency. http://www.ftc.gov/os/2013/02/130201mobileprivacyreport.pdf (2013). Accessed 24 Oct 2016
149. Canadian Offices of the Privacy Commissioners: Seizing opportunity: good privacy practices for developing mobile apps. http://www.priv.gc.ca/information/pub/gd_app_201210_e.pdf (2012). Accessed 24 Oct 2016
150. Harris, K.D.: Privacy on the go: recommendations for the mobile ecosystem. http://oag.ca.gov/sites/all/files/pdfs/privacy/privacy_on_the_go.pdf (2013). Accessed 24 Oct 2016
151. GSMA: User perspectives on mobile privacy. http://www.gsma.com/publicpolicy/wpcontent/uploads/2012/03/futuresightuserperspectivesonuserprivacy.pdf (2011). Accessed 24 Oct 2016
152. Sundsøy, P., Bjelland, J., Iqbal, A.M., Pentland, A.S., De Montjoye, Y.A.: Big data-driven marketing: how machine learning outperforms marketers' gut-feeling. In: Greenberg, A.M., Kennedy, W.G., Bos, N. (eds.) Social Computing, Behavioral-Cultural Modeling and Prediction. Springer, Berlin (2013)
153. Pasquale, F.: Reforming the law of reputation. Loyola Univ. Chicago Law J. **47**, 515–539 (2015)
154. Ombelet, P.J., Kuczerawy, A., Valcke, P.: Supervising automated journalists in the newsroom: liability for algorithmically produced news stories. Revue du Droit des Technologies de l'Information. http://papers.ssrn.com/sol3/papers.cfm?abstract_id=2768646 (forthcoming 2016). Accessed 24 Oct 2016
155. Latar, N.L., Norsfors, D.: Digital identities and journalism content—how artificial intelligence and journalism may co-develop and why society should care. Innov. Journalism. **6**(7), 1–47 (2006)
156. Ombelet, P.J., Morozov, E.: A robot stole my Pulitzer! How automated journalism and loss of reading privacy may hurt civil discourse. http://www. slate.com/articles/technology/future_tense/2012/03/narrative_science_robot_journalists_customized_news_and_the_danger_to_civil_ discourse_.single.html (2012). Accessed 24 Oct 2016
157. Hacker, P., Petkova, B.: Reining in the big promise of big data: transparency, inequality, and new regulatory frontiers. Northwest. J. Technol. Intellect. Prop. https://papers.ssrn.com/sol3/papers.cfm?abstract_id=2773527 (forthcoming 2016). Accessed 24 Oct 2016
158. Hajian, S., Domingo-Ferrer, J.: Direct and indirect discrimination prevention methods. In: Custers, B., Calders, T., Schermer, B., Zarsky, T. (eds.) Discrimination and Privacy in the Information Society. Springer, New York (2013)

159. Mayer-Schonberger, V., Cukier, K.: Big Data. A Revolution That Will Transform How We Live, Work, And Think. Eamon Dolan/Houghton Mifflin Harcourt, Boston, MA (2014)
160. Calders, T., Verwer, S.: Three naïve Bayes approaches for discrimination-free classification. Data Min. Knowl. Disc. **21**(2), 277–292 (2010)
161. Kamiran, F., Calders, T., Pechenizkiy, M.: Techniques for discrimination-free predictive models. In: Custers, B., Calders, T., Schermer, B., Zarsky, T. (eds.) Discrimination and Privacy in the Information Society. Springer, New York (2013)
162. Tutt, A.: An FDA for algorithms. Adm. Law Rev. **67**, 1–26 (2016)
163. FTC: Spring privacy series: alternative scoring products. http://www.ftc.gov/news-events/events-calendar/2014/03/spring-privacy-series-alternative-scoring-products (2014). Accessed 24 Oct 2016
164. Ramirez, E.: The privacy challenges of big data: a view from the lifeguard's chair. https://www.ftc.gov/public-statements/2013/08/privacy-challenges-big-data-view-lifeguard%E2%80%99s-chair (2013). Accessed 24 Oct 2016
165. Sandvig, C., Hamilton, K., Karahalios, K., Langbort, C.: Auditing algorithms: research methods for detecting discrimination on internet platforms. Data and discrimination: converting critical concerns into productive inquiry. http://www-personal.umich.edu/~csandvig/research/Auditing%20Algorithms%20--%20Sandvig%20--%20ICA%202014%20Data%20and%20Discrimination%20Preconference.pdf (2014). Accessed 24 Oct 2016
166. Benkler, Y.: The Wealth of Networks: How Social Production Transforms Markets and Freedom. Yale University Press, New Haven, CT (2006)
167. Citron, D.K.: Open code governance. Univ. Chicago Legal Forum. **2008**(1), 355–387 (2008)
168. Barnett, J.M.: The host's dilemma: strategic forfeiture in platform markets for informational goods. Harv. Law Rev. **124**(8), 1861–1938 (2011)
169. Moses, L.: Marketers should take note of when women feel least attractive: what messages to convey and when to send them. ADWEEK. http://www.adweek.com/news/advertising-branding/marketers-should-take-note-when-women-feel-least-attractive-152753 (2013). Accessed 24 Oct 2016
170. Orentlicher, D.: Prescription data mining and the protection of patients' interests. J. Law Med. Ethics. **38**(1), 74–84 (2010)
171. WPF: Data broker testimony results in new congressional letters to data brokers about vulnerability-based marketing. http://www.worldprivacyforum.org/2014/02/wpfs-data-broker-testimony-results-in-new-congressional-letters-to-data-brokers-regarding-vulnerability-based-marketing/ (2014). Accessed 24 Oct 2016
172. Bakos, Y., Marotta-Wurgler, F., Trossen, D.R.: Does anyone read the fine print? Consumer attention to standard-form contracts. J. Leg. Stud. **43**(1), 1–35 (2014)
173. MacDonald, A.M., Cranor, L.F.: The cost of reading privacy policies. J. Law Policy Inf. Soc. **4**(3), 540–565 (2008)
174. Lipford, H.R, Watson, J., Whitney, M., Froiland, K., Reeder, R.W.: Visual vs compact: a comparison of privacy policy interfaces. In: Proceedings of the SIGCHI Conference on Human Factors in Computing Systems, 1111–1114 (2010)
175. Passera, S., Haapio, H.: Transforming contracts from legal rules to user-centered communication tools: a human-information interaction challenge. Commun. Des. Q. Rev. **1**(3), 38–45 (2013)
176. Phillips, E.D.: The Software License Unveiled. Oxford University Press, Oxford (2009)
177. Gardner, T.: To read, or not to read... the terms and conditions. The Daily Mail. http://www.dailymail.co.uk/news/article-2118688/PayPalagreement-longer-Hamlet-iTunes-beats-Macbeth.html (2012). Accessed 24 Oct 2016
178. Ayres, I., Schwartz, A.: The no-reading problem in consumer contract law. Stanford Law Rev. **66**, 545 (2014)
179. Bar-Gill, O., Ben-Shahar, O.: Regulatory techniques in consumer protection: a critique of European consumer contract law. Common Mark. Law Rev. **50**, 109–126 (2013)

180. Luzak, J.: Passive consumers vs. the new online disclosure rules of the consumer rights. Directive. http://papers.ssrn.com/sol3/papers.cfm?abstract_id=2553877 (2014). Accessed 24 Oct 2016
181. Luzak, J.: To withdraw or not to withdraw? Evaluation of the mandatory right of withdrawal in consumer distance selling contracts taking into account its behavioral effects on consumers. J. Consum. Policy. **37**(1), 91–111 (2014)
182. Purnhagen, K., Van Herpen, E.: Can bonus packs mislead consumers? An empirical assessment of the ECJ's mars judgment and its potential impact on EU marketing regulation. In: Wageningen Working Papers Series in Law and Governance 2014/07, http://papers.ssrn.com/sol3/papers.cfm?abstract_id=2503342 (2014)
183. MacDonald, A.M., Reeder, R.W., Kelley, P.G., Cranor, L.F.: A comparative study of online privacy policies and formats. In: Goldberg, I., Atallah, M.J. (eds.) Privacy Enhancing Technologies. Springer, Berlin (2009)
184. Stigler, G.: The Economics of information. J. Polit. Econ. **69**(3), 213–225 (1961)
185. Akerlof, G.A.: The Market for "lemons": quality uncertainty and the market mechanisms. Q. J. Econ. **84**(3), 488 (1970)
186. Macho-Stadler, I., Pérez-Castrillo, J.D.: An Introduction to the Economics of Information. Oxford University Press, Oxford (2001)
187. Evans, M.B., McBride, A.A., Queen, M., Thayer, A., Spyridakis, J.H.: The effect of style and typography on perceptions of document tone. http://faculty.washington.edu/jansp/Publications/Document_Tone_IEEE_Proceedings_2004.pdf (2004). Accessed 24 Oct 2016
188. Masson, M.E.J., Waldron, M.A.: Comprehension of legal contracts by non-experts: effectiveness of plain language redrafting. Appl. Cogn. Psychol. **8**, 67–85 (1994)
189. Ben-Shahar, O., Schneider, C.E.: More Than You Wanted to Know: The Failure of Mandated Disclosure. Princeton University Press, Princeton (2014)
190. Radin, M.: Boilerplate. Princeton University Press, Princeton, NJ (2013)
191. Ben-Shahar, O., Chilton, A.S.: "Best practices" in the design of privacy disclosures: an experimental test. http://papers.ssrn.com/sol3/papers.cfm?abstract_id=2670115 (2015). Accessed 24 Oct 2016
192. Miller, A.A.: What do we worry about when we worry about price discrimination? The law and ethics of using personal information for pricing. J. Technol. Law Policy. **19**, 41–104 (2014)
193. Mittlestadt, B.D., Allo, P., Taddeo, M., Wachter, S., Floridi, L.: The ethics of algorithms: mapping the debate. Big Data Soc. 1–21 (2016)

AlgorithmWatch: What Role Can a Watchdog Organization Play in Ensuring Algorithmic Accountability?

Matthias Spielkamp

Abstract In early 2015, Nicholas Diakopoulos's paper "Algorithmic Accountability Reporting: On the Investigation of Black Boxes" sparked a debate in a small but international community of journalists, focusing on the question how journalists can contribute to the growing field of investigating automated decision making (ADM) systems and holding them accountable to democratic control. This started the process of a group of four people, consisting of a journalist, a data journalist, a data scientist and a philosopher, thinking about what kind of means were needed to increase public attention for this issue in Europe. It led to the creation of AlgorithmWatch, a watchdog and advocacy initiative based in Berlin. Its challenges are manyfold: to develop criteria as a basis for deciding what ADM processes to watch, develop criteria for the evaluation itself, come up with methods of how to do this, to find sources of funding for it, and more. This chapter provides first thoughts on how AlgorithmWatch will tackle these challenges, detailing its "ADM manifesto" and mission statement, and argues that there is a developing ecosystem of initiatives from different stakeholder groups in this rather new field of research and civil engagement.

Abbreviation

ADM Automated decision making

1 A Short History of Failures: How the Idea for AlgorithmWatch Came About

In the beginning of 2015, I had come across the idea of algorithmic accountability in the seminal paper by Nicholas Diakopoulos, "Algorithmic Accountability Reporting: On the Investigation of Black Boxes". Diakopoulos, a computer scientist by

M. Spielkamp (✉)
AlgorithmWatch, Oranienstr. 19a, 10999 Berlin, Germany
e-mail: ms@AlgorithmWatch.org

T. Cerquitelli et al. (eds.), *Transparent Data Mining for Big and Small Data*,
Studies in Big Data 32, DOI 10.1007/978-3-319-54024-5_9

training, had written it during his stay at the Tow Center for Digital Journalism at Columbia Journalism School, researching how to increase "clarity about how algorithms exercise their power over us" by using journalistic means. The paper sparked a debate in a small but international community of journalists.

In a nutshell, the basic question Diakopoulos asks in his paper is: Can journalistic reporting play a role in holding algorithmic decision making systems accountable to democratic control? This clearly is a loaded question, presupposing that there are such processes that need to be held accountable. For the purpose of this text, I assume that it does, as has been convincingly argued in a wide range of research, but will later come back to the importance of this question.

As a technophile journalist with a degree in philosophy, I had long worked on the intersection of technological change (mainly triggered by digitization and the Internet), legal regulation and ethical norm setting. My main focus was the debate about how societies needed to revamp their ideas of copyright, privacy and data security in the light of technological change. I had also reported on A.I. stories as early as 1997, when I produced a TV report on the chess tournament Deep Blue vs. Garry Kasparov. So it was quite clear to me immediately that Diakopoulos was on to something. It was not that clear, though, how successful and effective journalists could be with this endeavor, giving their generally technophobe attitude (in Germany) and the restraints of rapidly shrinking budgets.

My hope was that foundations would see the relevance of the issue and develop an interest in promoting early stage experiments. So I applied for funding at some of the very few German foundations dedicated to supporting journalism. The idea: Convene a workshop with a good mix of people from different backgrounds: journalism, computer science, law, and sociology. The goal: To find out what algorithmic accountability reporting should set its sights on, what qualifications were needed to pursue it, how such research would need to be organized, and what it could feasibly achieve.

The result: All potential funders found the idea very interesting, none eventually committed to provide even a meager amount of money for travel costs and an event location.

That was a huge disappointment, but what developing the workshop design and the question did was to entrench in my brain that there needs to be more thought put into this concept. And as Maslow's law of the instrument suggests, "if all you have is a hammer, everything looks like a nail". So when in May 2015 the German Volkswagen Foundation published a call for proposals titled "Science and Data-Driven Journalism", seeking to promote cooperation between research and data-driven journalism, I immediately saw an opportunity to apply the newly found hammer.

Volkswagen Foundation said by funding this program they were hoping that "the results of such collaboration would provide civil society with opportunities to overcome the challenges presented by 'Big Data'". This framing of the objective was rather unclear, as data journalism itself, albeit working with data—and sometimes 'big data'—does not inherently aim to focus on "challenges presented by 'Big Data'" but use data to look into all sorts of issues.

Half a data journalist myself, I had teamed up years before with one of Germany's preeminent data journalists, Lorenz Matzat, to train journalists in developing concepts for using data journalism in their newsrooms. I suggested to Lorenz we submit a proposal for a project, using data journalism tools and methods in order to implement a new approach, something that actually does focus on the challenges presented by big data: algorithmic accountability reporting. Although having been contacted by a number of other consortia who wanted to collaborate with him preparing proposals, he immediately agreed. The reason: He shared the assessment that this could really be a novel approach that would most likely benefit enormously from data visualization techniques used in data journalism.

The algorithmic process we set our sights on was predictive policing, the idea that with the help of automated data analytics, police can predict crime hotspots (or even potential offenders) and adjust their strategies accordingly. The reason we picked this issue was twofold: First, at the time, predictive policing experiments had just started in several German cities, secondly, it was apparent that the use of the technology could potentially have an impact on civil liberties, i.e. by creating no-go areas. Last not least: I had, in collaboration with a public research institute, submitted a proposal to look into predictive policing, to an international funder—again unsuccessfully, but it gave us a head start in terms of preparation.

In order to be able to investigate algorithmic decision making systems, we needed to collaborate with a computer scientist who would ideally share our assessment of the situation: that there is something that urgently needs exploration not just from a technical perspective but also from an ethical one. For the workshop I had planned earlier, Matzat had suggested as a participant Katharina Zweig, a professor for computer science (Informatik) at the Technical University of Kaiserslautern. Her website stated that she was an expert in graph theory and complex network analysis who had also created a program called "socioinformatics" at her university and stated that her research was focused on network analysis literacy. I had already invited her to a panel discussion I was hosting at the annual conference of investigative journalists in Hamburg and she had met the suggestion with enthusiasm to discuss ways to actually perform algorithmic accountability reporting in practice. She was also hooked immediately on the idea to submit a proposal to investigate predictive policing to Volkswagen Stiftung.

Developing this proposal together further sharpened the understanding we had of what this kind of reporting was supposed to achieve. As we wrote in the application:

> The goal [to investigate all steps of the long chain of responsibilities – from data collection and algorithm development to police action] is threefold. First, to investigate how these problems are communicated from the researcher - who invented a method - to the police who is acting on the predictions; second, to inform the public about the astonishingly problematic aspects of even the most simple forms of predictive policing; and third, to develop a guideline for a transparent communication of the intended scopes and limited applicability of statistical methods in big data analytics.

All of these aspects would later surface again in the discussions about the goals of AlgorithmWatch.

In September 2015, when the final decision was made, the person responsible for the selection procedure at Volkswagen Stiftung told us that our proposal was considered to be in ninth place of 84 submissions. Eight received funding.

A huge disappointment, but at the same time, the intensive and multi-disciplinary work on the proposal had fortified our view that the use of algorithmic decision making processes in many instances was in dire need of better observation, if not some kind of control.

During the next months, lots of emails were exchanged, discussing how to proceed. Was there another research program that we could tap to finance the predictive policing investigation? What other ideas could we develop in order to raise awareness about algorithmic accountability? What exactly was it that we, a tiny group of people with a mix of expertise, wanted to achieve anyway? Was it in fact time for a(nother) manifesto?

At the same time, all of us took part in these discussions already, by giving talks on the topic, following reporting on it or listening to other people present it at conferences and meetings. To us, it was more apparent than ever that there was a demand for a public dialogue on the subject, and therefore an opportunity to attract attention.

1.1 A Manifesto, a Mission Statement, and the Darn Ethics Thing

So it felt entirely logical when Lorenz Matzat one day told us that he had not just come up with the name AlgorithmWatch for our initiative but had also already registered it as a Twitter handle and a domain name (AlgorithmWatch.org). At the same time, we were ourselves hugely surprised by three things. First: that we had not come up with the idea to create a watchdog organization earlier. Second: that this was exactly what we thought was needed. And third: that no one else had thought of this before us. The name itself seemed to crystallize many ideas we had discussed during the month leading up to this but that we had failed to consolidate. We were electrified, but—as probably happens often in moments like this—we underestimated the bumpy road that lay ahead of us.

In early January of 2016, I wrote a lengthy email to my colleagues, presenting a plan of action, detailing how we should proceed in the coming months until the official launch of the AlgorithmWatch website, which would mark the start of the public life of our initiative. It was supposed to happen at re:publica, one of the world's largest conferences focused on digital culture, drawing thousands of participants to Berlin each year at the beginning of May. Until then, we wanted to be able to present at least two things: a manifesto and a mission statement.

We also decided to take on board another collaborator: Lorena Jaume-Palasí, a Berlin-based, Spanish-born researcher in philosophy, who is also a well-known expert in the fields of privacy and Internet governance. Having worked extensively on questions of what constitutes the public in the digital age, she seemed to be the ideal addition to our team's areas of expertise, especially when it came to fundamental questions of how to define what kinds of algorithmic decision making processes demand scrutiny.

Using funds of my Mercator Foundation fellowship I had available at the time, we organized a workshop in Berlin, convening the four of us for 2 days to exclusively work on a draft of the manifesto that we had developed in an online process. This was quite a strain, minding the fact that all of this work was in addition to our day jobs. But it was a watershed moment because for the first time it became clear to us what range of questions we were confronted with:

- What constitutes a decision (In legal terms? In ethical terms? In common sense terms?)
- What constitutes algorithmic decision making (ADM)?
- Who creates such processes?
- What kind of attitude do we have towards these processes?
- What demands do we voice, how can we justify them?
- What do we mean by regulation?
- How do we want to work as an organization?

In the end, we came up with version 1.0 of our ADM Manifesto:

Algorithmic decision making (ADM)* is a fact of life today; it will be a much bigger fact of life tomorrow. It carries enormous dangers; it holds enormous promise. The fact that most ADM procedures are black boxes to the people affected by them is not a law of nature. It must end.

ADM is never neutral.

- The creator of ADM is responsible for its results. ADM is created not only by its designer.
- ADM has to be intelligible in order to be held accountable to democratic control.
- Democratic societies have the duty to achieve intelligibility of ADM with a mix of technologies, regulation, and suitable oversight institutions.
- We have to decide how much of our freedom we allow ADM to preempt.

* We call the following process algorithmic decision making (ADM):

- design procedures to gather data,
- gather data,
- design algorithms to
- o analyse the data,
- o interpret the results of this analysis based on a human-defined interpretation model,
- o and to act automatically based on the interpretation as determined in a human-defined decision making model.

Besides this visible output, the main result of our discussions was the realization of two aspects that would be our hardest nuts to crack. One of them is a no-brainer: *How* would we actually watch ADM? Anyone dealing with this issue is well aware that there are enormous challenges in scrutinizing these complex systems. The other aspect, though, was not so evident: How would we justify *why* we want to watch certain ADM? Where would the criteria to make these decisions come from? More on these questions in the sections "What to watch?" and "How does AlgorithmWatch work, what can it achieve?"

1.2 The Launch

AlgorithmWatch launched with a presentation of the ideas behind the initiative and the website going live at re:publica on May 4, 2016. Being one of more than 750 talks on 17 stages at the conference, we managed to receive a tremendous amount of media attention, with four reports published on the first day, including articles on heise online and Golem.de, two of the most important German tech and policy news sites.[1] In addition, to our own surprise, within 10 days after the launch we were asked by two of the best-known mainstream media houses for expert opinion on issues revolving around algorithms and policy: RTL, Germany's largest private TV network, conducted an interview for their main evening news show, RTL aktuell, asking for an assessment of the allegations by a former Facebook news curator, that the curation team routinely suppressed or blacklisted topics of interest to conservatives.[2] And ZEIT Online requested comment on a joined statement by German federal ministries to a European Union Commission consultation, asking how far demands for making algorithms transparent should go. This was a clear indication our assumption was correct: There was a need for a focal point of the discussion revolving around algorithmic decision-making.

The number of invitations we received to conferences and workshops reinforced this impression. In the months following the launch (until September 2016) we were present at five national and international conferences, including Dataharvest in Mechelen (Belgium) and the European Dialogue on Internet Governance (EuroDIG) in Brussels, and several open and closed workshops, among them discussions hosted by the German Federal Ministries of the Interior and Economy. For the coming 6 months, there are at least another ten conference and workshop participations scheduled, both in Germany and abroad.

Feedback in these discussions was overwhelmingly positive, but questions remained the same throughout: How is AlgorithmWatch going to proceed, and what can a four-person charitable initiative achieve?

---

[1] http://algorithmwatch.org/aw-in-den-medien/.

[2] http://gizmodo.com/former-facebook-workers-we-routinely-suppressed-conser-1775461006.

1.3 What to Watch?

One of the main challenges we identified is how to determine what ADM to "watch". The public discussion often has an alarmist tone to it when it comes to the dangers of machines making decisions or preparing them, resulting in a bland critique of all algorithmic systems as beyond human control and adversary to fundamental rights. At the same time it is entirely obvious that many people have long been benefitting from automating decisions, be it by travelling in airplanes steered by auto pilots, by having their inboxes protected from a deluge of spam emails, or by enjoying climate control systems in their cars. None of them seem controversial to the general user, although at least the first two are intensely discussed and even controversial in the communities who develop these technologies (and, in addition, auto pilots are heavily regulated).

Also, the criteria to determine which ADM systems pose challenges for a democratic society cannot be drawn from law alone. Privacy and non-discrimination legislation can provide the grounds for demands for transparency and/or intelligibility in many cases. But what, for example, about the case of predictive policing, when an ADM system is not used to profile individuals but to create crime maps? No personalized data needs to be used to do this, so it was hardly a surprise that the Bavarian data protection supervisor consequentially gave a thumbs-up for the use of such methods in the cities of Nuremburg and Munich. But what about other effects the use of these systems might have—e.g. creating new crime hotspots by intensively paroling and therefore devaluating areas that only had a very small higher-than-average crime rate? What about Facebook's use of algorithms to determine what users see in their timeline—users who have entered into a contractual relationship with a private company? It must be clear that many of the most pressing questions posed by the use of ADM systems cannot be answered by resorting to applying existing law. They must be subjected to an intense public debate about the ethics of using these systems. (Thus item 5 of our manifesto: We have to decide how much of our freedom we allow ADM to preempt.)

Moreover, there are thousands, if not millions of ADM systems "out there"; no organization can ever look at all of them, no matter what resources it has available.

So neither the fact that some technology entails an ADM system in itself, nor the legal status of it, nor "general interest in an issue", can alone be a suitable guiding principle to determine what ADM to focus our attention on.

What is needed instead is a set of criteria that more precisely define the grounds for deciding what ADM are important enough to look at them more closely, something along the lines of "are fundamental rights challenged?", "is the public interest affected?" or similar.

As obvious as it seems, this is in fact hard to do. Because AlgorithmWatch is a trans-disciplinary initiative, we set out to do this in a parallel process: With the support of a fellowship by Bucerius Lab, a program created by ZEIT foundation promoting work that studies societal aspects of an increasingly digitized society, Lorena Jaume-Palasí drafted a discussion paper drawing up an outline for such a categorization of ADM. This work is still in progress, as the draft was discussed in a workshop in Mid-September with a group of experts from the fields of computer science, law, sociology, journalism and political sciences, as well as practitioners using ADM in their daily work, either individually or in their company. Publication is scheduled for December 2016, which will be the starting point of a wider discussion with communities concerned with algorithmic accountability and governance.

1.4 How Does AlgorithmWatch Work, What Can It Achieve?

Is it realistic to hope that a grass-roots initiative founded by four individuals in their spare time can wield influence in a field where the systems in question are applied by governments and global corporations? As with all new watchdog and advocacy organizations, this question will be decided by a mix of expertise, strategy, and funding.

AlgorithmWatch's mission is to focus on four activities, laid out in our mission statement:

Algorithm Watch Mission Statement

The more technology develops, the more complex it becomes. AlgorithmWatch believes that complexity must not mean incomprehensibility.

AlgorithmWatch is a non-profit initiative to evaluate and shed light on algorithmic decision making processes that have a social relevance, meaning they are used either to predict or prescribe human action or to make decisions automatically.

HOW DO WE WORK?

Watch

AlgorithmWatch analyses the effects of algorithmic decision making processes on human behaviour and points out ethical conflicts.

Explain

AlgorithmWatch explains the characteristics and effects of complex algorithmic decision making processes to a general public.

Network

AlgorithmWatch is a platform linking experts from different cultures and disciplines focused on the study of algorithmic decision making processes and their social impact.

Engage

In order to maximise the benefits of algorithmic decision-making processes for society, AlgorithmWatch assists in developing ideas and strategies to achieve intelligibility of these processes—with a mix of technologies, regulation, and suitable oversight institutions.

All of these activities are scalable, from once-in-a-while statements and articles we publish on our blog to white papers detailing complex regulatory proposals and full-blown investigations into advanced ADM systems, analyzing technological, legal and ethical aspects from development to deployment of a given system. At the same time and equally important, AlgorithmWatch can use a wide variety of different means to draw attention to ADM systems, from journalistic investigation to scientific research, from publishing editorials to consulting law makers on the issue.

This enables the organization to apply pressure in a variety of different means and thus avoid some of the traps in the discussion. One of the questions frequently asked of us is whether it is at all possible to achieve any kind of transparency or intelligibility of ADM systems. Most of the time, the assumption is that this can only be done by technological analyses of a certain algorithm or database (or both). This is very hard to do under present circumstances: corporations and state actors are granted extensive secrecy provisions, and because many technologies utilized are at the same time complex and fast-changing, analyses of the technological aspects would present a challenge to even the most sophisticated computer science researchers. So our strategy in such a case could be to point out these facts and pose the question whether it is desirable or acceptable to let technologies make decisions with far reaching consequences that cannot be intelligible to well-trained experts. This again points to the paramount importance of having criteria that allow a categorization of ADM systems into those that need to be watched and those that do not (and some in between) as described in the section "What to watch?"

1.5 AlgorithmWatch as Part of a Developing Ecosystem

Notions to create watchdog organizations are not born in a vacuum. What preceded the founding of AlgorithmWatch was a mounting criticism of the pervasiveness of ADM systems that were not subject to sufficient debate, let alone held accountable to democratic oversight. This criticism came from a wide range of stakeholders: scientists, journalists, activists, government representatives, and business people. Motivations vary from violations of fundamental rights and loss of state power to fears of being cheated on or facing competitive disadvantages. The approaches to tackle the issue vary accordingly: Conferences organized by scientists, activists, and governments, articles and books published by journalists and academics, competition probes triggered by complaints from companies or launched by governments, and regulatory proposals from any of the above.

What was missing, though, was the focal point for these discussions, the "single point of contact" for media, activists or regulators to turn to when faced with the issue. This is what AlgorithmWatch has set out to be. We will see in the coming years whether we can meet this objective.

Erratum to

Transparent Data Mining for Big and Small Data

Tania Cerquitelli, Daniele Quercia, and Frank Pasquale

T. Cerquitelli et al. (eds.), *Transparent Data Mining for Big and Small Data*,
Studies in Big Data 11, DOI 10.1007/978-3-319-54024-5

DOI 10.1007/978-3-319-54024-5_10

The published version of this book contains incorrect volume number as 11. The correct volume number for the book has been updated in both print and online version as 32.

The updated original online version for this book can be found at
http://dx.doi.org/10.1007/978-3-319-54024-5

T. Cerquitelli et al. (eds.), *Transparent Data Mining for Big and Small Data*,
Studies in Big Data 32, DOI 10.1007/978-3-319-54024-5_10